Advances in Neuroethics

Advances in neuroscience research are bringing to the forefront major benefits and ethical challenges for medicine and society. The ethical concerns related to patients with mental health and neurological conditions, as well as emerging social and philosophical problems created by advances in neuroscience, neurology and neurotechnology are addressed by a specialized and interdisciplinary field called neuroethics.

As neuroscience rapidly evolves, there is a need to define how society ought to move forward with respect to an ever growing range of issues. The ethical, legal and social ramifications of neuroscience, neurotechnology and neurology for research, patient care, and public health are diverse and far-reaching — and are only beginning to be understood.

In this context, the book series "Advances in Neuroethics" addresses how advances in brain sciences can be attended to for the benefit of patients and society at large.

More information about this series at http://www.springer.com/series/14360

Joé T. Martineau • Eric Racine
Editors

Organizational Neuroethics

Reflections on the Contributions of Neuroscience to Management Theories and Business Practices

Springer

Editors
Joé T. Martineau
Department of Management
HEC Montréal
Montreal, QC, Canada

Eric Racine
IRCM, Université de Montréal,
and McGill University
Montréal, QC, Canada

ISSN 2522-5677 ISSN 2522-5685 (electronic)
Advances in Neuroethics
ISBN 978-3-030-27176-3 ISBN 978-3-030-27177-0 (eBook)
https://doi.org/10.1007/978-3-030-27177-0

This Springer imprint is published by the registered company Springer Nature Switzerland AG.
The registered company address is: Gewerbestrasse 11, 6330 Cham, Switzerland

Acknowledgements

The editors are extremely grateful to all the contributors who have made this project possible by devoting their time to write the chapters of this volume. We would also like to thank the staff from the editorial office at Springer, and notably Ms. Madona Samuel, for their support throughout this project. Thanks to the co-editors of the series *Advances in Neuroethics*, Dr. Dubljević, Dr. Joterrand, and Dr. Jox for their support to this project. Special thanks to Ms. Asma Minyaoui for her supportive work on the manuscript. Joé T. Martineau acknowledges support from a postdoctoral award from the Fonds de Recherche du Québec—Société et Culture, and support from HEC Montréal. Eric Racine acknowledges support from a Fonds de Recherche du Québec—Santé senior scholar career award. Additionally, he would like to thank Stephanie Simpson and Corinne Lajoie of the Pragmatic Health Ethics Research Unit for their assistance in the preparation of the revisions and corrections for the contributing authors.

Contents

Introduction

1

Joé T. Martineau and Eric Racine

Contents

"Organizational neuroscience" is a new area of research that seeks to bridge neuroscience research, theories, and methods with management and organizational science. The goal of organizational neuroscience is to incorporate knowledge about brain processes underlying thoughts, behaviors, and attitudes of organizational actors in order to inform management theories and to assist management practice in understanding, predicting, and improving these behaviors in the workplace (Becker et al. 2011; Senior et al. 2011). Already, various examples of practical applications from organizational neuroscience research have emerged, such as the use of neurofeedback in order to modify behavior, notably for leadership skills development (Waldman et al. 2011); the development of consumer neuroscience research or neuromarketing techniques, for improved marketing and managerial practices (Hubert and Kenning 2008); or the use of cognitive or performance enhancement drugs (e.g., modafinil) to increase vigilance and manage fatigue of employees, especially in the military. Another recent area of organizational neuroscience research focuses on ethical decision-making and behavior of employees and

J. T. Martineau (✉)
Department of Management, HEC Montréal, Montreal, QC, Canada
e-mail: joe.trempe-martineau@hec.ca

E. Racine
IRCM, Montréal, QC, Canada

Université de Montréal, Montréal, QC, Canada

McGill University, Montréal, QC, Canada
e-mail: eric.racine@ircm.qc.ca

J. T. Martineau, E. Racine (eds.), *Organizational Neuroethics*, Advances in Neuroethics, https://doi.org/10.1007/978-3-030-27177-0_1

managers. However, organizational neuroscience research and applications sometimes face methodological and technological limitations, which have important implication for their responsible use. They also raise profound ethical challenges regarding, among others, the understanding and interpretation of neuroscience research and innovations, organizational responsibility, discrimination, well-being of employees, informed consent, and thus possible coercion and abuse on the part of organizations.

The idea that neuroscience has the potential to enrich and have fundamental implications for organizational studies, among others, is not entirely new, but rather part of an ongoing debate about the contribution of neuroscience to resolving societal and ethical issues, most of it taking place in the field of neuroethics (Racine 2010, 2013). At the boundary of neuroscience and ethics, neuroethics is an interdisciplinary field that aims to analyze and discuss the impact of the development of neuroscience and neurotechnology on different aspects of human life. This field is typically divided into two main areas of interest: (1) the ethics of neuroscience and (2) the neuroscience of ethics (Roskies 2002; Levy 2007). The ethics of neuroscience is mostly associated with bioethics and addresses ethical issues related to the development of research and clinical neuroscience. For example, this area deals with issues related to the determination of brain death, as well as those related to the use of neuroimaging techniques. In contrast, the neuroscience of ethics aims to integrate the results of neuroscience research to contribute to our understanding of ethics itself. For example, the use of functional neuroimaging techniques now allows us to better understand the mechanisms underlying moral reasoning and emotional processes such as empathy. In the neuroethics literature, celebrations of the promises and epistemic supremacy of neuroscience have been met with equally passionate anti-naturalist critiques of this view, as well as more moderate positions [for a review, see Racine (2010) and Pickersgill (2013)].

In this edited volume, we wish to introduce, define, and map a new "organizational neuroethics" as an emerging interdisciplinary field of research that reflects on the ethics of organizational neuroscience research and applications, as well as on the neuroscience of organizational ethics, drawing on the debates and reflections that have been taking place in the field of neuroethics in the past two decades. To our knowledge, *Organizational Neuroethics: Reflections on the Contributions of Neuroscience to Management Theories and Business Practices* is the first book to tackle this important topic and to do it in an encompassing way.

This book is divided in two parts, each one prefaced by a section introduction that provides a general overview of the topic and the contributions regrouped in the section. Part I, *The Ethics of Organizational Neuroscience*, provides an overview of the different ethical and methodological issues as well as the technological limitations associated with research and applications in organizational neuroscience, including topics such as the ethics of neuroleadership, consumer neuroscience research, and neuromarketing practices. The contributions featured also discuss the ethical issues related to the use of neurocognitive enhancement drugs in the workplace. Part II, *The Neuroscience of Organizational Ethics*, features contributions on topics such as the neuroscience of ethical, and unethical, decision-making in organizations, the neuroscience of online trust, and the neuroscience of empathy and its implications for organizations.

One important aspect of this volume is that it provides contributions from a real diversity of scholars, a diversity we sought and encouraged as editors. Emerging and well-established scholars coming from various fields such as business ethics, management, neuroscience, and neuroethics, as well as from different locations such as Canada, the Unites States, the United Kingdom, Denmark, Spain, and Japan, have generously contributed to this book. The reader will also find contradicting views between some of them, indicators of the debates going on in this new field. This diversity of point of views is also something we believe crucial to the deepening of our understanding of the ethical issues in organizational neuroscience.

The global objectives of this book are thus to paint a broad portrait of organizational neuroethics, to discuss its future, to separate wheat from chaff in matters of scientific promises, and to promote the development of organizational neuroscience in the most ethical way. Indeed, we believe that regardless of the interesting, seductive, and promising aspects of research and potential applications in organizational neuroscience, this cannot blind us and clear us from adopting a cautious stance toward the rapid development of this emerging field, as well as the important ethical reflection it engenders. We hope that this first book on the topic of organizational neuroethics will provide a balanced overview of organizational neuroscience research and its applications, appealing to a broad audience in management, business ethics, neuroscience, neuroethics, philosophy, and social sciences.

References

Becker WJ, Cropanzano R, Sanfey AG. Organizational neuroscience: taking organizational theory inside the neural black box. J Manag. 2011;37(4):933–61.

Hubert M, Kenning P. A current overview of consumer neuroscience. J Consum Behav. 2008;7 (4/5):272.

Levy N. Neuroethics: challenges for the 21st century. Cambridge: Cambridge University Press; 2007.

Pickersgill M. The social life of the brain: neuroscience in society. Curr Sociol. 2013;61(3):322–40.

Racine E. Pragmatic neuroethics. Improving treatment and understanding of the mind brain. Cambridge, MA: MIT Press; 2010.

Racine E. Pragmatism and the contribution of neuroscience to ethics. Essays Philis Human. 2013;21 (1):13–30.

Roskies A. Neuroethics for the new millennium. Neuron. 2002;35(1):21–3.

Senior C, Lee N, Butler M. Organizational cognitive neuroscience. Organ Sci. 2011;22(3):804–15.

Waldman DA, Balthazard PA, Peterson SJ. Leadership and neuroscience: can we revolutionize the way that inspirational leaders are identified and developed? Acad Manag Perspect. 2011;25 (1):60.

Part I

The Ethics of Organizational Neuroscience

Section Introduction: The Ethics of Organizational Neuroscience

2

Joé T. Martineau and Eric Racine

Organizational neuroscience is a recent area of research. Given the enthusiasm of researchers and organizations themselves, the main focus and efforts of the last 15 years have been invested in the development of research and its applications. Ethical as well as critical views on this research have remained marginal. We believe this first book on organizational neuroethics provides an opportunity to step back and reflect on the past and current organizational neuroscience research and applications, their promises, as well as the ethical issues and methodological limitations they raise. It is also an opportunity to reflect on the future of organizational neuroscience and on how it is possible to promote its development in the most ethical way and how to overcome identified limitations. This is the main topic of the first part of this book.

This first part features five chapters. Spence provides a critical analysis and compelling argument to tackle issues related to sensory marketing although he argues that neuromarketing per se may not be problematic. Kim and Waldman offer an overview of organizational neuroscience research on leadership and put forth that the issues raised by this use of neuroscience research may not be as vexing as it first appears. Lindebaum and colleagues review ethics content in the use of neuroscience in management. They find these discussions severely restricted, with limited engagement with the deeper humanistic issues at stake. Hubert and Hubert

J. T. Martineau (✉)
Department of Management, HEC Montréal, Montreal, QC, Canada
e-mail: joe.trempe-martineau@hec.ca

E. Racine
IRCM, Montréal, QC, Canada

Université de Montréal, Montréal, QC, Canada

McGill University, Montréal, QC, Canada
e-mail: eric.racine@ircm.qc.ca

J. T. Martineau, E. Racine (eds.), *Organizational Neuroethics*, Advances in Neuroethics, https://doi.org/10.1007/978-3-030-27177-0_2

propose a novel model to understand and propel discussions about the theoretical and methodological implications of consumer neuroscience. Finally, Dubljević and colleagues describe different pharmacological and device-based human performance enhancement technologies. They discuss ethical aspects of their use in the workplace with a focus on regulatory options available.

On the Ethics of Neuromarketing and Sensory Marketing

3

Charles Spence

Contents

Abstract

In this chapter, a number of key ethical issues associated with the recent emergence of the related fields of neuromarketing and sensory marketing are reviewed. Now that these new techniques are really starting to show their predictive mettle relative to other, more traditional, consumer psychology/behavioural testing approaches to marketing, questions around the ethics of stimulating the brain's "buy button" start to raise their head. Here, I want to question what exactly is so special, and so worrying, about "looking inside the mind of the consumer". I will argue that public fears around the dangers of neuromarketing have been overblown, at least up until the present time and, as far as I can see, for the foreseeable future. I do, though, want to raise a number of concerns around the growing influence of sensory marketing on our behaviour, focusing, in particular, on the world of food and drink marketing. Ultimately, I believe that the consumer of

C. Spence (✉)
Crossmodal Research Laboratory, Department of Experimental Psychology, Oxford University, Oxford, UK
e-mail: charles.spence@psy.ox.ac.uk

J. T. Martineau, E. Racine (eds.), *Organizational Neuroethics*, Advances in Neuroethics, https://doi.org/10.1007/978-3-030-27177-0_3

tomorrow may well have much to fear from the emerging neuroscience-inspired approaches to sensory marketing. In fact, before too long, we may all start to find ourselves being sensorially "nudged" into a range of less healthy food behaviours. As such, establishing clear ethical guidelines will, I believe, become an increasingly important issue for those working in the field.

3.1 Introduction

In recent years, there has been a rapid growth of interest in the fields of both neuromarketing (e.g. Ariely and Berns 2010; du Plessis 2011; Pradeep 2010; Renvoisé and Morin 2007) and sensory marketing (e.g. Cooper 2013; Hilton 2015; Hultén et al. 2009; Krishna 2013; Lindstrom 2005; Spinney 2013). Both are relatively new fields of research,[1] and both are linked by a shared desire to use the latest cognitive neuroscience insights and techniques in order to influence the behaviour of consumers. The press have certainly been very active in terms of their promotion of the dangers/potential of these new approaches to influencing/controlling consumer behaviour (e.g. Anonymous 2008; Bray 2007; Winnett n.d.; see also Racine et al. 2005, 2006; Racine et al. 2010). But, one has to ask, is the public concern with all this purported manipulation in the marketplace, elicited by press stories of marketers somehow targeting "the buy button" in the consumer's brain, really justified? The buy button is the name given to the region or, more likely, network of areas that influence the purchasing decision. As one commentator put it, "The ultimate goal is to identify a 'buy button' in the brain which can be targeted and triggered by future commercials" (Winnett n.d.; see also Witchalls 2004).[2] Or, one might counter, is public sentiment being stirred up needlessly by all the overexcited journalistic coverage (Lawton 2010; Lawton and Willis 2010)?[3] And, furthermore, one might also want to ask, just how different is the situation today from the mass hysteria that supposedly gripped so many North American consumers on the publication of Vance Packard's *The Hidden Persuaders* back in 1957?[4]

[1]According to Madan (2010), the term "neuromarketing" was first coined by Ale Smidts, in 2002, in an inaugural university address (Smidts 2002). That said, Anonymous (2004) suggests that the term was already in use by Gerry Zaltman in the closing years of the last century. The BrightHouse Institute, though, were the first to use the term in print (see Anonymous 2002).

[2]Though, as Blakeslee (2004) wryly notes in the title of her article: "If you have a 'buy button' in your brain, what pushes it?".

[3]Just take the conclusion from one newspaper article entitled "Admen seek 'buy button' in our brains" (Winnett n.d.): "A trip to the local supermarket or car showroom could soon become a psychological challenge with shoppers fighting subconscious urges that are manipulated by the marketers".

[4]Referencing the original, John Bunyard published *The Honest Persuaders* in 2010.

Some people are undoubtedly worried by the supposed power of neuromarketing. No wonder, when the neuromarketing companies themselves claim that they can "peer into the subconscious mind of the consumer" (quote from Wall 2013). Just take the following quote from Commercial Alert, a consumer group who consider neuromarketing both dangerous and Orwellian in scope: According to an article that appeared in *The New York Times*, Gary Ruskin, the group's executive director, wrote to the Senate Committee on Commerce, Science and Transportation—"What would happen in this country if corporate marketers and political consultants could literally peer inside our brains and chart the neural activity that leads to our selections in the supermarket and voting booth? What if they then could trigger this neural activity by various means, so as to modify our behavior to serve their own ends?" (quoted in Blakeslee 2004; see also Editorial 2004b; Grey et al. 2003). However, as Eric Spangenberg, at the time, dean of the College of Business at Washington State University, countered, "I don't think you are going to be able to make someone buy a car or a computer that they don't need, but you might persuade them to choose one model instead of another" (Anonymous 2008: 111). This is what Murphy et al. (2008: 297) term "a 'soft' attack on autonomy". Indeed, given the latest evidence from the field of sensory marketing suggesting that we can all be biased to choose and consume foods that our bodies most certainly don't need, then maybe we should now all be more worried about advances in the latter field. This is especially true when consumers are time and again shown to be seemingly completely unaware of the sensory factors ("nudges") that the research demonstrates really does influence their behaviour.

Worries about the dangers of neuromarketing have led some to take concrete action. France, for instance, has tried to crack down on the perceived rogue use of neuroscience. Back in 2011, the French parliament revised its 2004 rules on bioethics which were amended to read: "Brain-imaging methods can be used only for medical or scientific research purposes or in the context of court expertise". Of course, while this law effectively bans the commercial use of neuroimaging, neuromarketing companies need only cross the border to get around this legal restriction (see Oullier 2012).

3.2 What Is There to Worry About?

For a number of years, both the theoreticians and practitioners working in the field of neuromarketing have struggled to demonstrate the added benefit that their new neuroscience-inspired methods can provide over and above traditional consumer testing and focus group research (e.g. Catterall and Maclaran 2006; Lunt 1981; see Ariely and Berns (2010), for an authoritative state-of-the-art review from a few years ago). At the same time, commentators such as Raymond Tallis (2008) and Legrenzi and Umilta (2011) have vocally criticized what they characterize as a headlong rush by so many fields of study to embrace neuroimaging and explain all human phenomena in terms of brain activity. They have even coined a term, "neuromania", to describe this worrying trend. Indeed, according to an anonymous commentator writing in the top science journal *Nature Neuroscience* a little over a decade ago, "Neuromarketing is little more than a new fad exploited by scientists and marketing consultants to blind corporate clients with science" (Editorial 2004a; Laybourne and Lewis 2005).

It is certainly true that many people have questioned the validity of traditional methods of consumer testing, for instance, consumer panels and focus group research (see Spence 2009, for a review).[5] And, given the exceedingly high failure rates of new product launches documented in most categories, it is hard to disagree. Commentators typically suggest that somewhere in the region of 70–95% of all new products fail within a year of launch (Schneider and Hall 2011). The exact figure here depends on the commentator and the particular product category under consideration [see Spence (2016b), for a review; though see also Spence (2015b), on the dangers associated with taking such scientific-sounding, but typically unsubstantiated, percentages at face value]. However, while winning that battle has certainly not proved easy, there are now growing signs that some of the most innovative researchers working in the field of neuromarketing are finally starting to provide significant gains (e.g. see Berns and Moore 2012; Boksem and Smidts 2015; Falk et al. 2016; Kühn et al. 2016; Telpaz et al. 2015; Venkatraman et al. 2015, for some representative examples); typically, it has to be said they have done so by means of a combined methodologies approach—that is, by combining neuroimaging data with more traditional measures (see Jarrett 2015; Spence 2016a, b, for commentary and review).

It is at this point, then, that the question really starts to emerge (or become relevant) as to the ethics of this kind of approach to modifying consumer behaviour [see Nill and Schibrowsky (2007), for a more general review of ethics and marketing, and Hensel et al. (2017), for a practitioner perspective]. The situation, or so it can be argued, is now very different from what it was when Ariely and Berns wrote their landmark review of the field back in 2010 and had this to say of the state of the art: "It is not yet clear whether neuroimaging provides better data [than] other marketing methods" (Ariely and Berns 2010: 287).

In the sections that follow, I will take a clear look at the field of neuromarketing, followed by the emerging field of sensory marketing, and address some of the key ethical questions that are now being raised. However, to begin with, it is important to try and define our terms clearly. For, as I will argue below, the answer to the ethical question, certainly as far as it applies to the field of neuromarketing, hinges on the scope and breadth of one's definition. By contrast, defining sensory marketing would seem relatively uncontroversial. Indeed, I am not sure that anyone would disagree (at least not substantively) with Aradhna Krishna when she says that "sensory marketing can be defined as 'marketing that engages the consumers' senses and

---

[5] According to one industry commentator, "in focus groups, some people want to please, others to dominate—urges that can influence their choices. In interviews, consumers often say what they think the interviewer wants to hear" [Tim Partlin of Lieberman Research Worldwide, quoted in Blakeslee (2004: 5)].

affects their perception, judgment and behavior" (Krishna 2012: 332). By contrast, the scope and definition of neuromarketing seems to have broadened quite dramatically over recent years.

Ariely and Berns (2010: 284) define neuromarketing as "The application of neuroimaging methods to product marketing" (see Editorial 2004b; Fugate 2007, for similarly narrow definitions). Contrast this with the more recent definition provided by Stanton et al. (2016, p. 3). They state that "We consider neuromarketing to be the use of neuroscience and physiological research techniques to gain new insights into consumers' behavior, preferences, and decision making, as well as other aspects of human cognition and behavior related to marketing". The latter authors go on to say: "Neuromarketing seeks information and insights beyond that revealed by traditional techniques such as surveys, focus groups, experiments, and ethnography—with the goals of enhancing marketing theory and practice...".[6]

After having attended a recent World Neuromarketing Forum meeting in London (April 2017; http://www.neuromarketingworldforum.com/), my sense is that many of the practitioners in the field have now adopted this much broader definition. In fact, my fear is that pretty much any cognitive/experimental psychology paradigm is in danger of being incorporated into the bag of tricks that neuromarketing companies now consider their own. As such, it soon becomes much harder to try and draw a clear line between those techniques that fall under the neuromarketing umbrella and those that do not (the latter, I call "neuroscience-inspired", instead; Spence 2016a, b).

If the umbrella term of neuromarketing is restricted solely to indirect neuroimaging measures—here I am thinking of techniques such as functional magnetic resonance imaging (fMRI), electro-encephalography (EEG), magnetoencephalography (MEG), and near-infrared spectroscopy (NIRS)—then there is little reason for the consumer to be worried about the dangers. Indeed I, like a number of other commentators (see Stanton et al. 2016), believe that the potential of the approach has been oversold. However, if the definition of neuromarketing is broadened too much, then it feels like the real question becomes one of whether any neuroscience-inspired approach to marketing is ethical. The answer to the latter question has to be in the affirmative, as I would argue that there is really no clear qualitative dividing line in terms of the methods used, or the results obtained between the traditional methods of consumer research (that no one complained about) and the all-singing, all-dancing neuromarketing techniques that are being peddled these days.

[6] It is unclear quite what the "experiments" in the second sentence is really supposed to refer to here, given that both behavioural scientists and cognitive neuroscientists would, I imagine, insist that they conducted experiments. One could, I think, also legitimately claim that certain neuroimaging techniques, specifically event-related potentials, should actually be considered as traditional given that marketing and advertising researchers have been using them for almost half a century (e.g. Krugman 1971; Weinstein 1981; Weinstein et al. 1980).

Here, it is interesting to note how many of the neuromarketing companies[7] that have traditionally (I mean over the last couple of decades) stressed the benefits of indirectly recording the activity of the human brain are increasingly shifting their focus to more implicit behavioural testing methods instead, often using a variant of that most-sturdy of behavioural tasks, the Implicit Association Test (see Greenwald et al. 1998).[8] Corporate clients, it has to be said, are often lured in by the promise of the benefits of brain imaging, only to be palmed off once they have been hooked with some implicit behavioural or psychophysiological measures instead—think eye-tracking (e.g. Juravle et al. 2015; Piqueras-Fiszman et al. 2013), galvanic skin response (GSR) measurement,[9] heartbeat, pupillometry (pupil dilation), or facial micro-expression analysis. Once again, it is not that these latter techniques don't have value; rather it is their incorporation within the umbrella term of neuromarketing that surprises me.[10] In fact, it would seem that the very definition of the term "neuromarketing" is changing in light of the growing realization of the challenges that are associated with commercial neuroimaging for business. On this point, just take the contemporary definition of neuromarketing that we came across earlier.

3.2.1 Interim Summary

To summarize, I believe that there is little to worry about, ethically speaking, if one takes a narrow definition on neuromarketing because the scare stories regarding the usefulness of the approach have been oversold; see Stanton et al. (2016), for a similar conclusion. And, should one adopt a broader definition, then there would not appear to be anything "special" about the new neuroscience-inspired approaches to marketing (see Nill and Schibrowsky 2007, for a similar conclusion). I have to say that I, myself, favour the narrower, traditional definition.

[7]Of which there are a growing number: According to estimates, there are somewhere between 1 and 200 companies currently selling neuromarketing (see Plassmann et al. 2012; Wall 2013).

[8]Note that it is not that I see no use for the use of so-called "implicit" behavioural tests like the IAT, for business (e.g. Parise and Spence 2012). In fact, we have often used it over the years in our industry-funded research (Demattè et al. 2006, 2007b). It is just that I want to question their incorporation into the gamut of neuroimaging techniques.

[9]According to an analysis conducted by Fisher et al. (2010), GSR is a particularly popular product offering amongst many neuromarketing companies.

[10]As Chris Jarrett (2015) put it when describing a purported neuromarketing study claiming to be able to predict block-busters using heart rate, skin conductance, and breathing, "This wasn't genuine neuromarketing—it was really just wiring people up to some basic physiological measures to see how excited they were by a movie trailer".

3.3 Some Problematic Cases

The real concern in recent years has been with companies conducting experiments on customers without informing the latter that this is what they are doing. So, for example, a few years ago, Facebook started experimenting on its users in order to understand the phenomenon of social contagion (see Kramer et al. 2014). There was a public backlash when its users found out that they had been "experimented on" without their informed consent. Here, the issue was not really about what the company might have found out about the minds of their users but, rather, that the latter should have been informed that they were being "experimented on".[11] By contrast, there is little likelihood that those taking part in a neuroimaging study would not know about it (though see Tovino 2005).

In a much smaller example of the problems that we may increasingly come to face, the realization that they were being monitored also disturbed the customers when a digital advert crashed recently in an Oslo pizza store called Peppe's Pizza (Pettit 2017). The code that appeared where the advert should have been revealed that the customer's age, gender, and facial expression were being identified remotely via a small camera placed over the advert—again without the knowledge of customers. The newspaper covering the story described this as "dystopian face scanning", going on to say: "The Peppe's billboard was created by retail analytics company Kairos, Dinside reports. The company creates recognition software designed to help businesses 'convert more customers,' 'recognise, merchandise, monetize'" (Pettit 2017). According to the broad definition that we came across a little earlier, this would, I think, have to count as an example of ethically questionable neuromarketing. What is more, I suspect that this kind of remote monitoring of our attention/emotional state is starting to become much more common (see Knapton 2016; Lo 2016, for a couple of other recent examples), though, normally, it has to be said customers have been alerted to the technology's use in advance, thus side stepping, in my opinion, some of the more vexing ethical issues. Do we, though, as consumers, have the right to know when this is happening? This feels like the kind of area where ethical guidelines would be beneficial.

In terms of the dangers of neuromarketing, one could easily imagine, should the public be made aware of the potential power of these new marketing techniques, at least as all too often portrayed in, and by, the mass media, that their response would be similar to what happened 60 years ago. According to the media reports, something akin to mass hysteria followed the publication of Vance Packard's *The Hidden Persuaders* back in 1957. The latter best-selling expose documented the various

---

[11]Facebook teamed up with academic researchers in order to conduct a study in which they intentionally manipulated 689,000 users' mood states without their knowledge or consent. Specifically, the social network controlled the news feed of users over a 1-week period back in 2012. They carefully managed which emotional expressions Facebook users were exposed to. When the peer-reviewed, and ethics board-approved, study came out detailing the findings, the company was subject to a major public backlash. People were angered that the company had not acquired people's informed consent prior to their taking part in the study (see BBC 2014a, b).

ways in which émigré psychologists fleeing the Second World War in Europe had come to revolutionize the study of consumer behaviour through motivational research (Samuel 2010). Extensive mention was made of the psychoanalytic techniques used by colourful characters like Louis Cheskin and Ernest Dichter. The fear then, as now, was that the latest new techniques would give the marketers unprecedented insight into, and control over, the behaviour of consumers (Thompson 2003). The concern, in other words, is that advertising and marketing campaigns based on neuroimaging would become so effective that consumers would lose their ability to resist, that is, that their free will would be compromised, thus infringing their "personal privacy to a totally unacceptable degree" (Editorial 2004b; see also Wilson et al. 2008).

One has to ask the question of why neuromarketing should raise such concerns relative to other consumer psychology techniques. For, as Gemma Calvert, one of the founders of neuromarketing company Neurosense, noted some years ago: "This is a descriptive technique—it describes what the brain is doing. With fMRI you can't modify brain behavior. You can't make people go out and buy something" (Witchalls 2004). Is it the presumed power of such techniques to find the "buy button" in the human brain, or is there something sacrosanct about what goes on inside the head, that is at stake here? Is the fear here that the neuromarketing companies will bypass the consumer's mind by peering directly into the brain—in some sense bypassing their conscious control? Whether or not they were engaged in such practices, for a long time, many large companies were afraid of the consumer backlash should it become public knowledge that they had been engaged in neuromarketing research. And while that corporate concern for reputational damage does feel to have receded somewhat in recent years (see Anonymous 2008), the French government's banning of commercial neuromarketing back in 2011 (see Oullier 2012) has still provided some with the irresistible opportunity to deliberately discomfort the leaders of, e.g. McDonalds in France over their interest in the minds of their customers (see http://www.dailymotion.com/video/xr6cos_neuromarketing-votre-cerveau-les-interesse-1-2_news).

My personal impression, though, from a little over two decades working in the field, is that while we should undoubtedly be alert to the potential ethical concerns (see Eaton and Illes 2007), e.g. regarding the loss of autonomy in the marketplace and being nudged towards less-healthy food behaviours, consumers currently actually have little to fear from the field of commercial neuromarketing. On the one hand, this is because of the very practical limitation that the time required to plan, conduct, and analyse neuromarketing research, and, who knows, acquire the relevant ethical approval, typically falls outside of the timelines of most real-world marketing deadlines (see Spence 2016a, b, for a review). Here, though, we need to make an important distinction between "consumer neuroscience" and "commercial neuromarketing" (Javor et al. 2013; Plassmann et al. 2015).[12] The former term refers primarily to academic research on business-related issues; the latter, by contrast, primarily concerns the commercial use, and funding of, neuroimaging research to address business-relevant questions. It is more in the latter where the ethical concerns mostly reside.

The consumer neuroscience that appears in the peer-reviewed academic journals operates on much longer timelines than commercial neuromarketing. The researchers involved are also much more likely to have gone to the trouble of gaining ethical approval for their study in advance, most journals insisting that details of ethical approval are included in the final manuscript (see also Joffe 2014). In contrast to the more cautious approach represented by most consumer neuroscience, many of the claims made by those involved in commercial neuromarketing that have appeared in the press have been overblown. On occasion, those who have overstepped the mark in terms of what the data actually show (e.g. see Iacaboni et al. 2007) have elicited a fierce critical backlash from those academics worried about the reputational damage to the field of the misuse of neuroimaging data (Aron et al. 2007; Editorial 2007; see also Oullier 2012).[13] Another infamous example of a commercial neuromarketer getting it completely wrong also appeared as an Op-Ed in *The New York Times.* Marketer Martin Lindstrom suggested that just because the same part of the brain, namely the insular cortex, "lit up" when hearing the ring of an iPhone as in those studies of people who were actually in love, this meant that people literally "*loved their iPhone*" (emphasis in the original)! This, as has been pointed out by numerous commentators, is a particularly egregious example of a failure to understand the statistical properties of reverse inference (Herper 2011; Poldrack 2006, 2011).[14]

3.3.1 Some Outstanding Ethical Issues

Stanton et al. (2016) raise questions about the lack of ethical board oversight over much of the commercial neuromarketing research that is conducted.[15] I must admit

[12]According to Javor et al. (2013: 1), "We argue for a differentiated terminology, naming commercial applications of neuroscientific methods 'neuro-marketing' and scientific ones 'consumer neuroscience'".

[13]I have it on good authority that certain of the neuromarketers whose research was discussed in Lindstrom's (2008) book, *Buy-ology*, were less than happy with the way in which their scientific results were "twisted" to fit a particular marketing story!

[14]That said, given the replication crisis that is currently sweeping psychology (e.g. Francis 2014; Ioannidis et al. 2014; Open Science Collaboration 2015), not to mention many other disciplines (Ioannides 2005), one can legitimately question how many of the results that end up making it into the press are really robust.

[15]Though, as Brammer (2004) notes, the absence of any commercial fMRI machines in the UK means that all neuroimaging studies using this technique will likely have had to gain ethical approval prior to the start of data collection anyway. The situation is obviously very different when it comes to ERP research given the much cheaper cost of entry to the field.

that I don't really see this as much of a concern, but expanding on that point is undoubtedly a topic for another day (see Wagner 2003).[16] Certainly, many other researchers out there who conduct experiments would never think of applying for ethics. However, this is really a question about the ethics of seeking ethical approval, and not neuromarketing per se. Another ethical question that has been raised here concerns a lack of peer-review scrutiny over much of the commercial neuromarketing research that is carried out. Here, I agree with Wall's (2013) assessment that this is likely to lead to poor-quality science. However, I personally do not think that poor experimentation is in and of itself an ethical issue. Should the marketing manager wish to waste his/her budget on poor-quality scientific research, so be it![17] Others, meanwhile, have worried about the lack of disclosure when the authors of peer-reviewed academic papers fail to disclose their affiliations with the companies making the products being studied (see Gottwald 2017). This does indeed seem to be a murky area, and while a growing number of journals now explicitly require the authors to make a conflict of interest statement and state where the funding for the study came from, others still do not. So, one of the ethical issues here relates to the need to disclose the sources of funding that may be perceived by others (rightly or wrongly) as giving rise to a possible conflict of interest.

According to Murphy et al. (2008), one other area where ethical oversight might be needed is when neuromarketing is used to target vulnerable populations who may be harmed or exploited by the deployment of neuromarketing. While Murphy et al. have in mind those with some sort of impaired mental capacity (e.g. due to brain damage), one might wonder whether the clinically overweight and/or obese should perhaps be considered in the same category. If so, then there are serious ethical issues around the use of food and beverage-related sensory marketing. To my way of thinking, the real ethical challenge here relates to the fact that various aspects of the food environment would seem to be biasing us consumers towards less healthy food behaviours. At the same time, it is clear that most of us do not realize quite what impact these exogenous cues are having on us (i.e. we are all, in some sense, "vulnerable").

3.3.2 Multiple Uses for Commercial Neuromarketing

Where Stanton et al. (2016) are mistaken, I think, is in stating that there is real value in commercial neuromarketing only if the methods underpinning that research are valid. They state that "The goal of profit maximization might not lend itself thorough

[16]In fact, Wagner (2003: 23) starts his paper: "Few researchers at the beginning of the twenty-first century would submit a report of research involving human subjects without first obtaining review from an ethical review board".

[17]Research ethics wise, there could be issues in some countries where standards of research ethics imply that research will be of sufficient quality to justify any risks to research subjects (otherwise the risk is not justified). However, this depends on the underlying ethical code relied on in the country where the study is conducted.

(sic) scientific practice. Scientific results are worthwhile only if the methods used to collect the data are sound" (Stanton et al. 2016: 7). I would argue that such a view neglects the practical role of the latest expensive, "colourful" science (Barrera-Valencia 2015) in helping convince people and win internal arguments about strategy and the best approach within corporations. I would also say that the press/public fascination with neuroscience means the very connection between a company, or product/brand, and a neuroscience story is sufficient to have marketing appeal, pretty much regardless of what the evidence shows.

As a case in point, I would only point to my own group's neuroimaging research on the Lynx effect—Lynx, or Axe, as it is called outside the UK is one of the world's best-known brands. The underpinning science was open, peer-reviewed, and of the highest quality (I hasten to add). Nevertheless, the real value to the marketers funding the research was in gaining access to the magazines with a readership of relevant demographic (i.e. "lads' mags") many orders of magnitude higher than the journals in which we published the underpinning research (see Demattè et al. 2007a; McGlone et al. 2013). At the time, magazines like GQ, and Maxim, had a regular readership of young males in the relevant age group of something like two million every month. The neuroimaging study showed a shift in activation in the part of the orbitofrontal cortex coding facial attractiveness when a pleasant scent like Lynx (or for that matter the smell of roses) was presented rather than when an unpleasant smell of body odour or burnt rubber was delivered to the young female participant's noses. The latter had to rate the attractiveness of a carefully calibrated set of male faces leading eventually to a story about the parts of a woman's brain that men should be targeting.

3.3.3 Dealing with "the File Drawer Problem"

In the world of consumer neuroscience, much stock is placed in those studies where the researchers concerned have been able to use the latest neuroimaging analysis techniques, either in isolation or seemingly more frequently in combination with other methodologies/approaches, to predict some aspect of future consumer behaviour (e.g. see Berns and Moore 2012; Boksem and Smidts 2015; Falk et al. 2016; Kühn et al. 2016; Venkatraman et al. 2015). And while I certainly do not wish to cast doubt on the difficulty of obtaining such impressive-looking results, I do wonder about just how much weight is, or should be, placed on the findings. Note how these findings are typically published long after the predictions were made. Much more convincing, at least to my way of thinking, would be cases where predictions concerning future sales patterns, or consumer behaviour, are made public in advance of the relevant data being collected. The problem with the current state of affairs is that one never knows how many failed predictions are sitting in the filing cabinets [see Rosenthal (1979), for an early statement of the file drawer problem, and Simonsohn et al. (2014), for a possible means of assessing its prevalence in a given research field]. Now, while the file drawer problem is certainly not unique to the field of neuromarketing, I do worry that it may nevertheless hinder the uptake of these approaches in the minds of the marketing managers, whose jobs are on the line with seemingly every decision that they make. In fact, the evidence suggests that they are much more tempted to rely on "gut feel", as apparently they have always done (Blakeman 2017). And, according to an online survey of a little over 1000 marketers conducted back in 2014, 49% of them reported "trusting my gut" when it came to deciding where to invest their marketing budgets (Anonymous 2014).[18] Thus, even if the neuromarketers' predictions about the future success of the neuromarketing techniques that they employ were to be believed, that still does not necessarily mean that the marketing managers would take their findings/recommendations on board anyway.

3.3.4 Ethical Concerns with the Use of Sensory Marketing

Sensory marketing can be thought of as a *sensory* equivalent of the *informational* nudging championed by the behavioural economists over the last decade or so (see Thaler 2016; Thaler and Sunstein 2008). While some concerns have been voiced around the patrician undertones of many nudging interventions, it has to be said that it hasn't raised anything like the same level of public concern as has neuromarketing. It is not that people think that nudging is ineffective as a technique of behaviour change, but rather, given that it involves informational influence, I suppose the view is that we are all able consciously to make our own minds up about how to respond to such nudges, as when we are told how many of the previous guests in our hotel room chose to reuse their towels, say.[19] By contrast, there is, I think, a very real danger with sensory nudging that the influence on our behaviour may pass under the radar of consciousness as it were. Indeed, elsewhere, I have argued that a number of examples of sensory marketing, such as the red star on the front of the bottle of San Pellegrino, can be thought of as *functionally subliminal* in the sense that everyone knows it is there, but few realise what signals it may be sending to the customer's mind until they are told about it (see Spence 2012, for a review). It is also worth noting here that nudging, while potentially an effective means of getting us to buy more healthy food and beverage products, doesn't necessarily mean that we stop buying the unhealthy stuff. In effect, nudging can actually lead to increased food purchasing, at least under certain conditions (Kroese et al. 2016).[20]

[18]And the marketers are by no means exceptional in this regard: According to psychologist Gerd Gigerenzer, "I've worked with large companies and asked decision makers how often they base an important professional decision on that gut feeling. In the companies I've worked with, which are large international companies, about 50% of all decisions are at the end a gut decision" (quoted in Fox 2014).

[19]Here, it may be that the majority of the published examples of nudging would seem to be targeted at societally desirable behaviours, such as donating our organs or saving more for our retirements. By contrast, much of the discussion around neuromarketing has focused on corporate greed.

[20]Kroese et al. (2016) conducted a study in three snack shops located in train stations. They assessed whether the customers could be nudged towards healthier food choices simply by moving the fruit

3.3.5 Sensory Marketing of Foods in an Obesogenic Environment

In the West, we are all being exposed to ever more food-related marketing, what some have chosen to call "gastroporn" or "food porn" (see Spence et al. 2016, for a review). One of the dangers here is that the highly calorific and unhealthy ingredient-laden recipes we see set consumption norms that are inappropriate. So, even if we do not necessarily make the dishes themselves, and the evidence suggests that we don't, we may nevertheless still internalize an inappropriate norm which guides our consumption behaviour in other situations. However, I really do wonder whether a greater concern may relate to olfactory, rather than visual, marketing. Here I am thinking of the release of food-related aromas that may provoke us to consume more than we otherwise might. No laughing matter in this era of the growing obesity crisis spreading across so many countries (see Spence 2015a, for a review). Intriguingly, some of the latest marketing research has highlighted the potential of ambient food-related aromas to increase appetite (Zoon et al. 2016), induce salivation (Proserpio et al. 2017), and even bias consumer's food choice (Gaillet et al. 2013; see also Trivedi 2006). So there is a very real danger that the environments in which we all live and work are becoming more obesogenic as a result of the "intelligent" use of the latest findings and insights from the field of sensory marketing (e.g. from the targeted release of those odours that signify highly energy-dense foods), in particular. This would seem ethically questionable given consumers' lack of awareness about how environmental sensory cues affect them. That said, I think that we are still quite some way from realizing the dangers that Ariely and Berns (2010: 289) pointed to when they imagined a future in which companies might use neuromarketing to create "a 'super-heroin of food'—a product so delicious that all but the most ascetic individuals would find it irresistible".[21]

There is undoubtedly growing interest in the use of olfactory cues to help grow sales, even when that scent is not intrinsic to the food or beverage product(s) being sold (Hilton 2015; Nassauer 2014; Spence 2015a). To illustrate the potential concern, just take the findings of one recent study conducted in a Dutch supermarket showing that pumping out the scent of melon into the aisles led to a 14% increase in sales (Leenders et al. 2019; though see de Wijk et al. 2018, for contradictory results). According to the journalist reporting on the study, these "findings could help supermarkets in attempts to make us spend more" (Howarth 2016). It certainly

---

closer to the check-out till than the snacks—the reverse normally being the case. The "nudge" worked in the sense that people were indeed more likely to buy a piece of fruit or a muesli bar. That said, those customers whose behaviour was assessed continued to purchase snacks like crisps, cookies, and chocolate *as well*. In other words, an intervention that was designed to *cut* people's consumption may actually have resulted in their purchasing *more* calories instead.

[21] In passing, one might also consider "the free lunch" as but another long-standing example of sensory marketing. Once again, the ethics of this particular appeal to the senses are questionable, at least to the extent that it is assumed to influence decision-making via the stomach rather than the rational mind (see Anonymous 1972; Farrell 1965; Halvorson and Rudelius 1977, for an early discussion of the topic), which the latest evidence suggests it most definitely does (see Spence 2016a, b, for a review).

Table 3.1 Number (and % in brackets) of bottles of French vs. German wine sold as a function of the type of background music played in North et al.'s (1997) study

Background Music

	French accordian music	German Bierkeller music
Bottles of French wine sold	40 (77%)	12 (23%)
Bottles of German wine sold	8 (27%)	22 (73%)

Graph reprinted from Spence (2012; Physiol Behav; 107:505–15. https://doi.org/10.1016/j.p)

gives a whole new meaning to the popular phrase "paying through the nose"! To date, it is surprising how few complaints such approaches have attracted, other than by those suffering from adverse response to certain olfactory stimuli (Byatt 2003; Spence 2002).[22]

This lack of public concern is somewhat surprising given long-standing suggestions that there are, for example, certain unnamed scents that may encourage people to linger for longer at the slot machines in the casinos (see Hirsch 1995). What is also noticeable about much of the sensory marketing research is how little awareness many consumers seem to have that the environmental sensory cues are actually biasing their behaviour, despite the sales data to the contrary (see Spence et al. 2014, for a review). Just take North et al.'s (1997, 1999) study of the impact of playing either French or German music on wine sales in a British supermarket. The pattern of sales showed a *dramatic* reversal as a function of the music that was being played in the background (see Table 3.1). That said, only 6 out of the 44 consumers who were quizzed after they left the tills admitted that the background music had had any impact on their purchasing behaviour! And while the shoppers were presumably aware of the music playing in the background, evidence from the field of olfaction suggests that scents sometimes have a larger impact on our perception/behaviour when we aren't aware of their presence (e.g. Li et al. 2007; see also Gaillet-Torrent et al. 2014). It would, I think, be great to see a replication of North et al.'s study, given the most impressive and potentially important findings reported therein. Certainly, the evidence from a broad array of studies now concurs in showing that

[22]One of the only examples comes from the Milk Advertising Board in California who were forced to remove their cookie-scented adverts from bus shelters because it was deemed insensitive to homeless people who might use the shelters to sleep in (Cuneo 2006). Meanwhile, a couple of decades ago, attempts to scent the London Underground were also met by hostility from Londoners, and the scheme was soon abandoned too (see Spence 2002). These findings suggest that one can go too far with specifically olfactory marketing.

ambient sensory cues can, and often do, exert a significant effect on consumer behaviour (see Spence et al. 2014, for a review), though it has to be said there is nothing out there yet that is quite as impressive as North et al.'s oft-cited findings (albeit based on total sales of little more than 100 bottles of wine).

3.4 Conclusion

As this review of the burgeoning literature on neuromarketing and sensory marketing has hopefully made clear, I believe that there is little to worry about the ethics of neuromarketing to date. This is, in part, because of its limited practical usefulness in the marketplace. One certainly has to take press coverage of the latest findings in this area with a hefty grain of salt. Any hysteria that there has been has most often resulted from unrealistic hype around the potential of the technique (O'Connor et al. 2012; Stanton et al. 2016). It is certainly true that the emergence of more sophisticated neuroimaging data analysis techniques, especially when integrated into a combined methodologies approach, is now starting to show some real promise in terms of predicting large-scale future consumer behaviour, often on the basis of relatively small sample sizes (e.g. see Boksem and Smidts 2015; Falk et al. 2016; Kühn et al. 2016; Venkatraman et al. 2012; though see also Venkatraman et al. 2015). That said, the evidence still largely suggests that the majority of marketing decisions are made on gut feel, not hard data.

I would argue that a much more worrying trend[23] is the rise in sensory marketing, which tends to be inspired by the neuroscience, even if not always relying on such techniques directly (this is what I call the "neuroscience-inspired" approach). Appealing to the senses, in ever more effective ways, especially in the world of food and drink marketing is producing food landscapes that may provoke increased consumption of larger unhealthier portions than ever before. As yet, the dangers associated with such sensory marketing approaches have gone largely unremarked upon. That said, there is a growing body of evidence that such forms of marketing are increasingly nudging us into less than healthy food behaviours (see Spence 2015a, 2017; Spence et al. 2016, for reviews). I believe, given the evidence demonstrating that people are unaware how their behaviour is being manipulated by such sensory cues, that there are important ethical questions to be addressed here centred around the loss of autonomy/free will.

Now, even if the latest neuromarketing and sensory marketing techniques could do everything that the press would have us believe that they can, that still doesn't guarantee that these techniques will necessarily influence the decisions we make in the marketplace. The reason is that the marketing managers apparently still primarily

[23]Worrying, both because of the seeming effectiveness of these kinds of interventions and because of most consumers' lack of awareness that exogenous factors are biasing their decisions/behaviour. As a case in point, just think about the increasingly obesogenic environment in which the increasingly obese populace finds itself.

make their decisions based on "gut" feel (Anonymous 2014; cf. Blakeman 2017). Furthermore, the behaviours and intentions of the participants in the laboratory do not always lead to an appropriate prediction of what will happen out in the wild (e.g. see Forwood et al. 2015). When your job is on the line, there is always a worry about how generalizable the neuromarketing findings really are! The file drawer problem is somewhat less relevant in the case of sensory marketing, where the results of any intervention are often measured in terms of sales data, anyway.

In conclusion, some of the most challenging ethical cases to have emerged in the world of neuromarketing and sensory would seem to be when people are experimented on without their informed consent having been sought. As the cases reported in this review have hopefully made clear, this is starting to become an increasingly worrying problem. As yet, there does not appear to be any clear guidance about whether customers need to be informed in advance. If the intervention is classed as an experiment, then the likelihood is yes, but if instead it is merely labelled a responsive marketing intervention, then I think the situation is far less clear.

Disclosure Statement Over the last two decades, Prof. Spence has consulted with many of the world's largest companies on the topics of neuromarketing and sensory marketing. He has published psychophysical and neuromarketing studies of the Lynx effect on behalf of Unilever. This research is referred to in this chapter.

References

Anonymous. Does breaking bread break the rules? Purchasing 1972 Oct 24;57.

Anonymous. Brighthouse institute for thought sciences launches first "neuromarketing" research company. 2002 Jun 22. http://www.prweb.com/releases/2002/6/prweb40936.php.

Anonymous. Inside the mind of the consumer. The Economist 2004 Jun 10. http://www.economist.com/node/2724481.

Anonymous. The way the brain buys. The Economist 2008 Dec 20;109–11.

Anonymous. Digital roadblock: marketers struggle to reinvent themselves. 2014. https://www.adobe.com/content/dam/Adobe/en/solutions/digital-marketing/pdfs/adobe-digital-roadblock-survey.pdf.

Ariely D, Berns GS. Neuromarketing: the hope and hype of neuroimaging in business. Nat Rev Neurosci. 2010;11:284–92. https://doi.org/10.1038/nrn2795.

Aron A, Badre D, Brett M, Cacioppo J, Chambers C, Cools R, et al. Politics and the brain (letter to the editor). The New York Times 2007 Nov 14. http://www.nytimes.com/2007/11/14/opinion/lweb14brain.html.

Barrera-Valencia M. Editorial. Neurotechnologies: the need for an ethical commitment in their implementation. Revista ces Psicología. 2015;8(2 Julio–Diciembre):i–iii.

BBC Facebook emotion experiment sparks criticism. BBC News 2014a Jun 30. http://www.bbc.com/news/technology-28051930.

BBC Facebook admits failings over emotion manipulation study. BBC News 2014b Oct 3. http://www.bbc.com/news/technology-29475019.

Berns GS, Moore S. A neural predictor of cultural popularity. J Consum Psychol. 2012;22:154–60. https://doi.org/10.1016/j.jcps.2011.05.001.

Blakeman S. Why 'gut feel' is better than 'Big Data'. Pulse 2017 Mar 20. https://www.linkedin.com/pulse/why-gut-feel-better-than-big-data-steve-blakeman.

Blakeslee S. If your brain has a 'Buy Button,' in your brain, what pushes it? The New York Times 2004 Oct 19;(154):5. http://www.nytimes.com/2004/10/19/science/if-your-brain-has-a-buybutton-what-pushes-it.html?_r=0.

Boksem MAS, Smidts A. Brain responses to movie trailers predict individual preferences for movies and their population-wide commercial success. J Mark Res. 2015;52:482–92. https://doi.org/10.1509/jmr.13.0572.

Brammer M. Brain scam? Nat Neurosci. 2004;7:1015. https://doi.org/10.1038/nn1004-1015.

Bray P. Switched on to buy. The Telegraph Magazine 2007 Feb 17;4.

Bunyard J. The honest persuaders. London: Lulu Enterprises; 2010.

Byatt AS. How we lost our sense of smell. In: Bragg M, James PD, editors. Sightlines. London: Vintage Books; 2003. p. 257–67.

Catterall M, Maclaran P. Focus groups in marketing research. In: Belk RW, editor. Handbook of qualitative research methods in marketing. Cheltenham: Edward Elgar; 2006. p. 255–67.

Cooper L. Sensory marketing—could it be worth $100m to brands? Marketing Week 2013 Oct 31. http://www.marketingweek.com/2013/10/30/sensory-marketing-could-it-be-worth-100m-to-brands/.

Cuneo AC. Milk board forced to remove outdoor scent strip ads. AdAge 2006 Dec 6. http://adage.com/article/news/milkboard-forced-remove-outdoor-scent-strip-ads/113643/.

de Wijk RA, Maaskant A, Kremer S, Holthuysen N, Stijnen D. Supermarket shopper movements versus sales, and the effects of scent, light and sound. Food Qual Prefer. 2018;70:32–9. https://doi.org/10.1016/j.foodqual.2017.03.010.

Demattè ML, Sanabria D, Spence C. Cross-modal associations between odors and colors. Chem Senses. 2006;31:531–8. https://doi.org/10.1093/chemse/bjj057.

Demattè ML, Österbauer R, Spence C. Olfactory cues modulate judgments of facial attractiveness. Chem Senses. 2007a;32:603–10. https://doi.org/10.1093/chemse/bjm030.

Demattè ML, Sanabria D, Spence C. Olfactory-tactile compatibility effects demonstrated using the implicit association task. Acta Psychol. 2007b;124:332–43. https://doi.org/10.1016/j.actpsy.2006.04.001.

du Plessis E. The branded mind: what neuroscience really tells us about the puzzle of the brain and the brand. London: KoganPage; 2011.

Eaton ML, Illes J. Commercializing cognitive neurotechnology—the ethical terrain. Nat Biotechnol. 2007;25:393–7.

Editorial. Mind games. Nat Neurosci. 2004a;7:683.

Editorial. Neuromarketing: beyond branding. Lancet Neurol. 2004b;3:71.

Editorial. Mind games. Nature. 2007;450:457. https://doi.org/10.1038/450457a.

Falk EB, O'Donnell MB, Tompson S, Gonzalez R, Cin SD, Strecher V, et al. Functional brain imaging predicts public health campaign success. Soc Cogn Affect Neurosci. 2016;11:204–14. https://doi.org/10.1093/scan/nsv108.

Farrell PV. Eats, ethics, and economics. Purchasing 1965 Dec 2;5.

Fisher CE, Chin L, Klitzman R. Defining neuromarketing: practices and professional challenges. Harv Rev Psychiatry. 2010;18:230–7. https://doi.org/10.3109/10673229.2010.496623.

Forwood SE, Ahern AL, Hollands GJ, Ng YL, Marteau TM. Priming healthy eating: you can't prime all the people all of the time. Appetite. 2015;89:93–102. https://doi.org/10.1016/j.appet.2015.01.018.

Fox J. Instinct can beat analytical thinking. Harv Bus Rev 2014:20. https://hbr.org/2014/06/instinct-can-beat-analytical-thinking.

Francis G. The frequency of excess success for articles in Psychological Science. Psychon Bull Rev. 2014;21:1180–7. https://doi.org/10.3758/s13423-014-0601-x.

Fugate DL. Neuromarketing: a layman's look at neuroscience and its potential application to marketing practice. J Consum Mark. 2007;24:385–94.

Gaillet M, Sulmont-Rossé C, Issanchou S, Chabanet C, Chambaron S. Priming effects of an olfactory food cue on subsequent food-related behaviour. Food Qual Prefer. 2013;30:274–81.

Gaillet-Torrent M, Sulmont-Rossé C, Issanchou S, Chabanet C, Chambaron S. Impact of a non-attentively perceived odour on subsequent food choices. Appetite. 2014;76:17–22.
Gottwald J. Can brain imaging predict chocolate sales? In search of the buy button. BPS Digest 2017 Feb 8. https://digest.bps.org.uk/2017/02/08/can-brain-activity-predict-chocolate-sales-in-search-of-the-buy-button/.
Greenwald AG, McGhee DE, Schwartz JLK. Measuring individual differences in implicit cognition: the implicit association test. J Pers Soc Psychol. 1998;74:1464–80.
Grey T, Healy JM, Linn S, Rowe J, Ruskin G, Villani VS. Commercial alert asks Emory University to halt neuromarketing experiments. 2003 [Cited in Stanton et al. 2016].
Halvorson PJ, Rudelius W. Is there a free lunch? Assessing the luncheon meeting between sales representatives and buyers. J Mark. 1977:44–9.
Hensel D, Iorga A, Wolter L, Znanewitz J. Conducting neuromarketing studies ethically: practitioner perspectives. Cogent Psychol. 2017;4:1320858. https://doi.org/10.1080/23311908.2017.1320858.
Herper M. No, you don't love your iPhone in that way. Forbes 2011 Oct 2. https://www.forbes.com/sites/matthewherper/2011/10/02/no-you-dont-love-your-iphone-in-that-way/#6cc02128636c.
Hilton K. The science of sensory marketing. Harv Bus Rev. 2015:28–31.
Hirsch AR. Effects of ambient odors on slot-machine usage in a Las Vegas casino. Psychol Mark. 1995;12:585–94.
Howarth M. We spend more if there's a fruity smell in the air: how one supermarket saw sales soar when it pumped out the scent of melon. DailyMail Online 2016 Jun 24. http://www.dailymail.co.uk/news/article-3658830/We-spend-s-fruity-smell-air-one-supermarket-saw-sales-soar-pumped-scent-melon.html.
Hultén B, Broweus N, van Dijk M. Sensory marketing. Basingstoke: Palgrave Macmillan; 2009.
Iacaboni M, Freedman J, Kaplan J, Jamieson KH, Freedman T, Knapp B, Fitzgerald K. This is your brain on politics. The New York Times 2007 Nov 11. http://www.nytimes.com/2007/11/11/opinion/11freedman.html?pagewanted=all&_r=0.
Ioannides JPA. Why most published research findings are false. PLoS Med. 2005;2(8):e124, 0696-0701. https://doi.org/10.1371/journal.pmed.0020124.
Ioannidis JPA, Munafò MR, Fusar-Poli P, Nosek BA, David SP. Publication and other reporting biases in cognitive sciences: detection, prevalence, and prevention. Trends Cogn Sci. 2014;18:235–41. https://doi.org/10.1016/j.tics.2014.02.010.
Jarrett C. Has the age of neuromarketing finally arrived? New York Magazine (Science of Us) 2015 Oct 14. http://nymag.com/scienceofus/2015/10/has-the-age-of-neuromarketing-finally-arrived.html.
Javor A, Koller M, Lee N, Chamberlain L, Ransmayr G. Neuromarketing and consumer neuroscience: contributions to neurology. BMC Neurol. 2013;13:13. https://doi.org/10.1186/1471-2377-13-13.
Joffe S. Revolution or reform in human subjects research oversight. J Law Med Ethics. 2014;40:922–9.
Juravle G, Velasco C, Salgado-Montejo A, Spence C. The hand grasps the centre, while the eyes saccade to the top of novel objects. Front Psychol. 2015;6:633. https://doi.org/10.3389/fpsyg.2015.00633.
Knapton S. Bad mood? New app senses emotion and suggests food to lift spirits. Telegraph 2016 Nov 27. http://www.telegraph.co.uk/science/2016/11/27/bad-mood-new-app-senses-emotion-suggests-food-lift-spirits/.
Kramer ADI, Guillory JE, Hancock JT. Experimental evidence of massive-scale emotional contagion through social networks. PNAS 2014;111:8788–90. http://www.pnas.org/content/111/24/8788.full.pdf.
Krishna A. An integrative review of sensory marketing: engaging the senses to affect perception, judgment and behavior. J Consum Psychol. 2012;22:332–51. https://doi.org/10.1016/j.jcps.2011.08.003.

Krishna A. Customer sense: how the 5 senses influence buying behaviour. New York: Palgrave Macmillan; 2013.
Kroese FM, Marchiori DR, de Ridder DTD. Nudging healthy food choices: a field experiment at the train station. J Public Health. 2016;38:e133–7. https://doi.org/10.1093/pubmed/fdv096.
Krugman HE. Brain wave measures of media involvement. J Advert Res. 1971;11:3–9.
Kühn S, Strelow E, Gallinat J. Multiple "buy buttons" in the brain: forecasting chocolate sales at point-of-sale based on functional brain activation using fMRI. NeuroImage. 2016;136:122–8. https://doi.org/10.1016/j.neuroimage.2016.05.021.
Lawton G. Neuromarketing: the results are in. New Scientist 2010 Sept 7. https://www.newscientist.com/article/dn19412-neuromarketing-the-results-are-in/.
Lawton G, Willis C. Mind-reading marketers have ways of making you buy. New Scientist 2010 Aug 4. https://www.newscientist.com/article/mg20727721-300-mind-reading-marketers-have-ways-of-making-you-buy/.
Laybourne P, Lewis D. Neuromarketing: the future of consumer research? Admap. 2005;461:28–30.
Leenders MAAM, Smidts A, El Haji A. Ambient scent as a mood inducer in supermarkets: the role of scent intensity and time-pressure of shoppers. J Retail Consum Serv. 2019;48:270–80. https://doi.org/10.1016/j.jretconser.2016.05.007.
Legrenzi P, Umilta C. Neuromania: on the limits of brain science (translated by F. Anderson). Oxford: Oxford University Press; 2011.
Li W, Moallem I, Paller KA, Gottfried JA. Subliminal smells can guide social preferences. Psychol Sci. 2007;18:1044–9. https://doi.org/10.1111/j.1467-9280.2007.02023.x.
Lindstrom M. Brand sense: how to build brands through touch, taste, smell, sight and sound. London: Kogan Page; 2005.
Lindstrom M. Buy-ology: how everything we believe about why we buy is wrong. London: Random House Business Books; 2008.
Lindstrom M. You love your iPhone. Literally. The New York Times 2011 Sept 30. http://www.nytimes.com/2011/10/01/opinion/you-love-your-iphone-literally.html.
Lo T. Finger LOOKIN' good: diners at Chinese KFC restaurant can use facial recognition technology to choose meals for them based on their mood and age. Daily Mail Online 2016 Dec 30. http://www.dailymail.co.uk/news/peoplesdaily/article-4073172/Do-LOOK-like-want-chicken-wings-KFC-uses-facial-recognition-predict-customers-orders-based-appearance-mood.html.
Lunt SG. Using focus groups in packaging research. In: Stern W, editor. Handbook of package design research. New York: Wiley Interscience; 1981. p. 112–24.
Madan CR. Neuromarketing: the next step in marketing research? Eureka. 2010;1(1):34–42.
McGlone F, Österbauer RA, Demattè ML, Spence C. The crossmodal influence of odor hedonics on facial attractiveness: behavioural and fMRI measures. In: Brain mapping. Rijeka: InTech; 2013. p. 209–25.
Murphy ER, Illes J, Reiner PB. Neuroethics of neuromarketing. J Consum Behav. 2008;7:293–302. https://doi.org/10.1002/cb.252.
Nassauer S. Using scent as a marketing tool, stores hope it—and shoppers—will linger: how Cinnabon, Lush Cosmetics, Panera Bread regulate smells in stores to get you to spend more. The Wall Street Journal 2014 May 20. http://www.wsj.com/articles/SB10001424052702303468704579573953132979382.
Nill A, Schibrowsky JA. Research on marketing ethics: a systematic review of the literature. J Macromark. 2007;27:256–73.
North AC, Hargreaves DJ, McKendrick J. In-store music affects product choice. Nature. 1997;390:132. https://doi.org/10.1038/36484.
North AC, Hargreaves DJ, McKendrick J. The influence of in-store music on wine selections. J Appl Psychol. 1999;84:271–6.
O'Connor C, Rees G, Joffe H. Neuroscience in the public sphere. Neuron. 2012;74:220–6. https://doi.org/10.1016/j.neuron.2012.04.004.

Open Science Collaboration. Estimating the reproducibility of psychological science. Science. 2015;349:943, aac4716-1–aac4716-8. https://doi.org/10.1126/science.aac4716.
Oullier O. Clear up this fuzzy thinking on brain scans. Nature. 2012;483:7. https://doi.org/10.1038/483007a.
Packard V. The hidden persuaders. Harmondsworth. Middlesex: Penguin Books; 1957.
Parise CV, Spence C. Assessing the associations between brand packaging and brand attributes using an indirect performance measure. Food Qual Prefer. 2012;24:17–23. https://doi.org/10.1016/j.foodqual.2011.08.004.
Pettit H. Creepy pizza sign is caught using a hidden camera to scan customers for their facial expressions and gender. Daily Mail Online, 2017 May 11. http://www.dailymail.co.uk/sciencetech/article-4495782/Pizza-sign-caught-using-hidden-camera-scan-customers.html.
Piqueras-Fiszman B, Velasco C, Salgado-Montejo A, Spence C. Combined eye tracking and word association analysis to evaluate the impact of changing the multisensory attributes of food packaging. Food Qual Prefer. 2013;28:328–38. https://doi.org/10.1016/j.foodqual.2012.10.006.
Plassmann H, Ramsøy TZ, Milosavljevic M. Branding the brain: a critical review and outlook. J Consum Psychol. 2012;22:18–36. https://doi.org/10.1016/j.jcps.2011.11.010.
Plassmann H, Venkatraman V, Huettel S, Yoon C. Consumer neuroscience: applications, challenges, and possible solutions. J Mark Res. 2015;52:427–35. https://doi.org/10.1509/jmr.14.0048.
Poldrack RA. Can cognitive processes be inferred from neuroimaging data? Trends Cogn Sci. 2006;10:59–63. https://doi.org/10.1016/j.tics.2005.12.004.
Poldrack RA. Inferring mental states from neuroimaging data: from reverse inference to large-scale decoding. Neuron. 2011;72:692–7. https://doi.org/10.1016/j.neuron.2011.11.001.
Pradeep AK. The buying brain: Secrets of selling to the subconscious mind. Hoboken: Wiley; 2010.
Proserpio C, de Graaf C, Laureati M, Pagliarini E, Boesveldt S. Impact of ambient odors on food intake, saliva production and appetite ratings. Physiol Behav. 2017;174:35–41. https://doi.org/10.1016/j.physbeh.2017.02.042.
Racine E, Bar-Ilan O, Illes J. fMRI in the public eye. Nat Rev Neurosci. 2005;6:159–64. https://doi.org/10.1038/nrn1609.
Racine E, Bar-Ilan O, Illes J. Brain imaging: a decade of coverage in the print media. Sci Commun. 2006;28:122–42.
Racine E, Waldman S, Rosenberg J, Illes J. Contemporary neuroscience in the media. Soc Sci Med. 2010;71:725–33. https://doi.org/10.1016/j.socscimed.2010.05.017.
Renvoisé P, Morin C. Neuromarketing: understanding the “buy buttons” in your customer’s brain. Nashville, TN: Thomas Nelson; 2007.
Rosenthal R. The “file drawer problem” and tolerance for null results. Psychol Bull. 1979;86:638–41.
Samuel LR. Freud on Madison Avenue: motivation research and subliminal advertising in America. Oxford: University of Pennsylvania Press; 2010.
Schneider J, Hall J. Why most product launches fail. Harv Bus Rev. 2011;89:21–3.
Simonsohn U, Nelson LD, Simmons JP. P-curve: a key to the file-drawer. J Exp Psychol. 2014;143:534–47. https://doi.org/10.1037/a0033242.
Smidts A. Kijken in het brein: Over de mogelijkheden van neuromarketing. (Brain imaging: Opportunities for neuromarketing). Inaugural Address Erasmus University: ERIM EIA-12-MKT. 2002.
Spence C. The ICI report on the secret of the senses. London: The Communication Group; 2002.
Spence C. Measuring the impossible. In: MINET Conference: measurement, sensation and cognition. Teddington: National Physical Laboratories; 2009. p. 53–61. isbn:978-0-946754-56-4.
Spence C. Managing sensory expectations concerning products and brands: capitalizing on the potential of sound and shape symbolism. J Consum Psychol. 2012;22:37–54. https://doi.org/10.1016/j.jcps.2011.09.004.
Spence C. Leading the consumer by the nose: on the commercialization of olfactory-design for the food and beverage sector. Flavour. 2015a;4:31. https://doi.org/10.1186/s13411-015-0041-1.

Spence C. Just how much of what we taste derives from the sense of smell? Flavour. 2015b;4:30. https://doi.org/10.1186/s13411-015-0040-2.
Spence C. Gastrodiplomacy: assessing the role of food in decision-making. Flavour. 2016a;5:4. https://doi.org/10.1186/s13411-016-0050-8.
Spence C. Neuroscience-inspired design: from academic neuromarketing to commercially-relevant research. Organ Res Meth. 2016b; https://doi.org/10.1177/1094428116672003.
Spence C. Gastrophysics: the new science of eating. London: Viking Penguin; 2017.
Spence C, Puccinelli N, Grewal D, Roggeveen AL. Store atmospherics: a multisensory perspective. Psychol Mark. 2014;31:472–88. https://doi.org/10.1002/mar.20709.
Spence C, Okajima K, Cheok AD, Petit O, Michel C. Eating with our eyes: from visual hunger to digital satiation. Brain Cogn. 2016;110:53–63. https://doi.org/10.1016/j.bandc.2015.08.006.
Spinney L. Selling sensation: the new marketing territory. New Scientist 2013 Sept 18;2934. https://www.newscientist.com/article/mg21929340-400-selling-sensation-the-new-marketing-territory/.
Stanton SJ, Sinnott-Armstrong W, Huettel S. Neuromarketing: ethical implications of its use and potential misuse. J Bus Ethics. 2016; https://doi.org/10.1007/s10551-016-3059-0.
Tallis R. The neuroscience delusion. Times Lit Supple 2008 Apr 9.
Telpaz A, Webb R, Levy DJ. Using EEG to predict consumers' future choices. J Mark Res. 2015;52:511–29. https://doi.org/10.1509/jmr.13.0564.
Thaler RH. Misbehaving. London: Allen Lane; 2016.
Thaler RH, Sunstein CR. Nudge: improving decisions about health, wealth and happiness. London: Penguin; 2008.
Thompson C. There's a sucker born in every medial prefrontal cortex. N Y Times Magazine 2003 Oct 26. http://www.nytimes.com/2003/10/26/magazine/there-s-a-sucker-born-in-every-medial-prefrontal-cortex.html.
Tovino SA. The confidentiality and privacy implications of functional magnetic resonance imaging. J Law Med Ethics. 2005;33:844–50.
Trivedi B. Recruiting smell for the hard sell. New Sci, 2006;2582:36–39. https://www.newscientist.com/article/mg19225821-800-recruiting-smell-for-the-hard-sell/.
Venkatraman V, Clithero JA, Fitzsimons GJ, Huellel S. New scanner data for brand marketers: how neuroscience can help better understand differences in brand preferences. J Consum Psychol. 2012;22:143–53. https://doi.org/10.1016/j.jcps.2011.11.008.
Venkatraman V, Dimoka A, Pavlou PA, Vo K, Hampton W, Bollinger B, et al. Predicting advertising success beyond traditional measures: new insights from neurophysiological methods and market response modeling. J Mark Res. 2015;52:436–52. https://doi.org/10.1509/jmr.13.0593.
Wagner RM. Ethical review of research involving human subjects: when and why is IRB review necessary? Muscle Nerve. 2003;28:27–39.
Wall M. What are neuromarketers really selling? The poor data and shoddy logic behind a hyped business boom. Slate 2013 July 16. http://www.slate.com/articles/health_and_science/science/2013/07/does_neuromarketing_work_poor_data_secret_analysis_and_logical_errors.html.
Weinstein S. Brain wave analysis: the beginning and future of package design research. In: Stern W, editor. Handbook of package design research. New York: Wiley Interscience; 1981. p. 492–504.
Weinstein S, Appel V, Weinstein C. Brain-activity responses to magazine and television advertising. J Adv Res. 1980;20:57–63.
Wilson RM, Gaines J, Hill RP. Neuromarketing and consumer free will. J Consum Affairs. 2008;42:389–410.
Winnett R. Admen seek 'buy button' in our brains. Unattributed English broadsheet newspaper article from c. 2003. n.d.
Witchalls C. Pushing the buy button. Newsweek 2004 Mar 21. http://www.newsweek.com/pushing-buy-button-123737.
Zoon HFA, de Graaf C, Boesveldt S. Food odours direct specific appetite. Foods. 2016;5:12. https://doi.org/10.3390/foods5010012.

Neuroethics in Leadership Research and Practice

4

Joohyung Kim and David A. Waldman

Contents

Abstract

In recent years, there has been a great proliferation of interest in neuroscience applications to both research and practice in the area of leadership. In terms of research, this interest raises questions regarding the treatment of research participants, storage and use of data, and so forth. But even greater issues have arisen with regard to practice, including areas pertaining to employee selection or placement, neurofeedback, and what might be termed "brain training." Despite valid concerns, we argue that ethical considerations in the neuroscience of leadership may not be as vexing as they appear. Instead, these considerations can be readily addressed, particularly with regard to basic research methods. On the other hand, continuing attention will need to be paid to practice-based, ethical issues.

J. Kim · D. A. Waldman (✉)
W. P. Carey School of Business, Arizona State University, Tempe, AZ, USA
e-mail: Jenny.J.Kim@asu.edu; waldman@asu.edu

J. T. Martineau, E. Racine (eds.), *Organizational Neuroethics*, Advances in Neuroethics, https://doi.org/10.1007/978-3-030-27177-0_4

4.1 Introduction

Imagine that as part of your job application process, you were told that the company uses applicants' brain profiles as part of its evaluation criteria, although your participation in the process is completely voluntary. The Human Resource manager further explains that the company has recently adopted a new brain training program for employees based on what she terms "neurofeedback," the participation in which is also at your discretion. Would you willingly be part of the process to have the company measure your brain activity and train your brain in a way the company desires? If not, what makes you feel reluctant?

In recent years, neuroscientific techniques have contributed to leadership research. By pinpointing the underlying neurological patterns of effective leaders, organizational neuroscience has shown the potential to enhance leadership research. Specifically, several studies have shown that the electrical patterns of a brain in a resting state, which reflects a relatively stable dispositional characteristic, explain unique variance in leadership phenomena and outcomes above and beyond traditional survey data (Balthazard et al. 2012; Hannah et al. 2013; Waldman et al. 2017a). These findings of distinct neural patterns have indicated that incorporating neural data can improve prediction.

However, along with such promising utility of new technologies, potential ethical concerns have been consistently raised. With the emergence of the field of neuroethics (Marcus 2002), neuroscientists, philosophers, and social scientists have begun to discuss ethical/moral issues with respect to the research and practice of applied neuroscience. More recently, organizational neuroscience (ON) scholars have also raised similar concerns, particularly those that are relevant to employment settings. For example, Waldman et al. (2015) cautioned against the use of neuroscreening and brain enhancement techniques in personnel selection and training, respectively. They specifically showed concern about the misuse of ON-based technologies as a means of discrimination and coercion by employers. These issues have been predominantly raised in leadership research, since the results of research, especially research based on the brain at rest (i.e., brain activities in a relaxed setting without external stimuli), could potentially be translated into practice. In other words, since ethical issues apply not only to study participants but also potentially to a wider range of employees in organizations, neuroethics in leadership can evoke a number of potential controversies among researchers and Human Resource (HR) practitioners.

While the most vexing ethical issues pertain to the realm of practice, it is also important to consider the relevance of such issues to basic academic research. We argue that many of the issues pertaining to the *neuroscience* aspects of some recent research often arise from a misunderstanding of core neurosensing or neuroimaging techniques. Individuals with little experience with neuroscience-based methods might be especially prone to false claims regarding their validity and ethical basis. In this chapter, we attempt to address the misconceptions that have been largely expressed by such individuals, while simultaneously considering ethical issues that have real substance.

We begin our chapter by discussing the utility of neuroleadership research. We provide research evidence on the role of neuroscience in leadership and other related fields and on the potential of leadership interventions. Then, we point out some caveats in incorporating neuroscience into traditional organizational research and provide general ethical guidelines to deal with future ethical issues. Finally, we identify, from a practical standpoint, the important legal and ethical aspects of using these innovative methods and technologies in employee selection, placement, and training.

Although some individuals have argued that management scholars adopting neuroscientific methodologies tend to neglect ethical issues (e.g., Lindebaum 2016), we argue that such claims are exaggerated. Scholars in this burgeoning area have indeed cautioned against immoral or unethical uses of neuroscience-based methods (e.g., Waldman et al. 2015). We agree that the potential for misuse, especially with regard to practical applications, is real. Nevertheless, we purport that open and fact-based discussions that accurately consider neuroscience methods and technologies and their potential uses (or misuses) may alleviate ethical concerns pertaining to both research and practice.

4.2 The Neuroscience of Leadership

Organizational scholars have only recently turned their attention to neuroscience concepts and methods. While disciplines such as economics and marketing embraced neuroscientific approaches about 20 years ago (Glimcher et al. 2009; Sebastian 2014), organizational neuroscience has a relatively short history. This shorter history may be attributed to three reasons: (1) differences in fundamental approaches, (2) technological issues, and (3) study samples. First, the predominant view of neuroscience has been based on reactivity of the human brain, as a response to external stimuli (Braeutigam et al. 2017; Raichle 2010). Such a perspective was suitable for economics or marketing where the main interest is in how individuals process given information and make decisions. However, the laboratory-based, stimulus-response approach was less suitable for organizational research, where subtle, complex social contexts have a large impact on employee behaviors.

In addition, functional magnetic resonance imaging (fMRI) technology has gained popularity relative to other comparable technologies across disciplines (Jack et al. 2017). Although fMRI has better spatial resolution than other technologies such as electroencephalography (EEG), its costs and technical sophistication (i.e., required know-how) are also more prohibitive, which has largely discouraged organizational scholars from adopting fMRI. Lastly, the field of organizational behavior has often discouraged the use of undergraduate students as a viable substitute for study sample, in the place of actual organizational members, due to the lack of external validity (Dobbins et al. 1988; Highhouse 2009). This puts a burden on organizational researchers to recruit working adults who are willing to spend significant time in participating in studies using neuroscience technologies, especially a technique like fMRI that involves confined space and potential discomfort.

It was not until recently that neuroscience scholars have increasingly adopted an intrinsic view of the human brain (see explanation below) to assess stable, trait-like proclivity (Biswal et al. 2010; Cole et al. 2014). That is, as opposed to recognizing changes in brain activities that are induced by stimuli (e.g., through the use of fMRI), researchers have come to realize the value of studying baseline, ongoing neural structures when the brain is at rest. This new trend, along with the advancement of quantitative EEG (qEEG) technology, has contributed to more cost-effective and expedient data collection processes. That is, the lack of expense, efficient time, and relative ease of qEEG measurement allows for more practical assessment. In addition, qEEG, as it accurately captures "ongoing spontaneous electrical activity" (Raichle 2010: 181), has demonstrated its ability to further an understanding of organizational phenomena, especially leadership.

Specifically, studies have shown the potential of neuroscience technologies in identifying effective leaders. For example, Balthazard et al. (2012) were able to distinguish transformational from non-transformational leaders based on qEEG variables with 92.5% accuracy. In addition, past findings have revealed that neurological variables can explain unique variance in outcomes above and beyond the traditional psychometric properties. For instance, researchers were able to predict leaders' adaptive performance with leader self-complexity measured through both psychological and neurological assessment (Hannah et al. 2013). Interestingly, not only were the two distinct measures correlated, but also each uniquely predicted leaders' decision-making performance, enabling a more comprehensive understanding of leader behaviors. Likewise, Waldman and colleagues (2017a) showed how followers' perceptions of ethical leadership, as well as moral relativism, were predicted by leaders' neurological profiles, particularly in the brain's default mode network (DMN).

Along with these findings in the neuroscience of leadership, research from related disciplines is also noteworthy. For instance, there is a growing body of research identifying brain regions associated with personality, adopting methodologies from neuroscience and relevant fields including genetics, physiology, and pharmacology (DeYoung et al. 2010). Scholars have identified distinct neurological profiles for Machiavellianism (Verbeke et al. 2011), trait anxiety (Kim and Whalen 2009), and Big Five personality (DeYoung et al. 2010). The emergence of social cognitive neuroscience has also led to an increasing scholarly interest in the underlying neural mechanisms of individuals' understanding of self and others (Lieberman 2007; Singer and Lamm 2009). The accumulation of such findings can potentially contribute to leadership research, since both personality traits and interpersonal interactions are major components of effective leadership.

Furthermore, the extended application of neuro-training programs, which are designed to enhance cognitive functions of the brain through computerized programs, to a healthy population implies the utility of neuroscientific applications to leadership development. One example of such applications is neurofeedback, a noninvasive procedure in which participants can control and adapt their brain activity in real time. Although it has been predominantly used for clinical purposes, the number of studies on its use for healthy individuals has increased over the past few years. Such training programs have been shown to be effective in improving individuals' attention, memory, intelligence, mood, and well-being (Gruzelier 2014).

More related to leadership development, studies have also shown the efficacy of neurofeedback sessions in improving the ability to regulate one's emotions (Johnston et al. 2010; Zotev et al. 2013) and activating neural patterns associated with certain emotions (Moll et al. 2014). Specifically, in Moll et al.'s (2014) study, participants were randomly assigned to either a neurofeedback or a control condition and were instructed to recall their past experiences of feeling affection toward others, pride, and neutral emotions and relive each emotion. While reliving the emotion, participants observed an image of a distorted ring and were told different meanings depending on the condition. Specifically, in the neurofeedback condition, the extent of distortion reflected the relative proximity of their real-time brain signals to the targeted emotion, while in the control condition, the distortion was random. After a few sessions, those in the neurofeedback condition learned to activate the neural patterns associated with the target emotion, while those in the control condition did not. We will consider neurofeedback in greater detail later in this chapter.

In short, research adopting neuroscience has shown its potential role in improving our understanding of leadership. Further, findings pertaining to neurofeedback have demonstrated interesting possibilities. However, it is still an open question regarding the extent to which neurofeedback might be relevant to leader development efforts. In particular, since the efficacy of neurofeedback interventions has been shown in relatively simple tasks (Gruzelier 2014), more thorough investigation is encouraged to assess its potential for broader leadership development, which necessitates complex decision making and interpersonal skills.

4.3 Ethical Issues in Neuroleadership Research

Most of the major ethical concerns with regard to organizational neuroscience research arise from two broad issues: (1) methodological concerns and (2) researchers' ethical decisions. The former has to do with the validity of the neuroscientific technology and its underlying assumptions, while the latter is concerned with discretionary actions of the researchers using the technology. Thus, in the following paragraphs, we first discuss the basic procedures, potential benefits, and risks of major neuroscientific methodologies, which can help reduce the inordinate fear about neurosensing and neuroimaging technologies. Then, turning to the second issue, we discuss specific dilemmas that researchers may encounter and provide ethical principles that can guide them to resolve such dilemmas, based on the authors' direct experiences and prior recommendations for handling such issues.

4.3.1 Methodological Concerns

Although neuroscientific technologies have been successfully utilized in various disciplines, organizational scholars are still skeptical about the validity and potential risks of these technologies. One may doubt whether there is sufficient validity to draw conclusions from data obtained from such sources (e.g., Lindebaum 2016). In addition, authors have raised questions about the potential harm of neurosensing or neuroimaging technologies on participants, especially when combined with neurofeedback training (e.g., Lindebaum and Raftopoulou 2017). However, these concerns generally arise from the lack of a clear understanding of such technologies. Therefore, in the following paragraphs, we first describe two major technologies and the underlying assumptions of two distinct research streams and subsequently assess their validity.

Technical Issues Functional magnetic resonance imaging (fMRI) and quantitative electroencephalography (qEEG) are two major technologies that have been used in cognitive neuroscience (Logothetis 2008). fMRI provides graphical representations of activated brain regions, primarily inferred from changes in the concentration of oxygen in the blood (blood-oxygen-level-dependent (BOLD) contrast; Logothetis 2008). On the other hand, EEG deals with electrical energy in the brain in terms of the amplitude and frequency of electrical signals across categorized bands (e.g., theta, alpha, beta, gamma). qEEG is an automated version of the traditional EEG, in which computer programs systematically analyze electrical brain activity, rather than having clinical experts manually scrutinize charts of brainwave activity (Thakor and Tong 2004).

Using fMRI and qEEG methods, participants are lying down in a brain scanner or wearing an EEG cap on their heads, respectively, throughout assessment. In addition, since brain signals are sensitive to head movement, eye movement, and other noises, participants are instructed to minimize their physical movements. When using fMRI, experimental stimuli are often given in an oral manner (e.g., audio or video), instead of a written format, to minimize potential influences of artifacts (e.g., Bagozzi et al. 2013). Moreover, since brain signals on a single task may interfere with another task, there is usually a time lag between tasks until the brain returns to the resting state, especially in experiments designed to assess within-person differences.

As for the technological issues, both fMRI and qEEG have their own advantages and disadvantages that can complement each other (Waldman et al. 2017b). Functional MRI has a better spatial resolution than qEEG and thus can better pinpoint brain activities in a more sophisticated manner. On the other hand, qEEG has a better temporal resolution, since fMRI involves a time lag between signals in the brain and recording of the corresponding neurophysiological change (e.g., BOLD signal; Raichle 2010). Furthermore, the recordings of both fMRI and qEEG often vary with physiological states (Thakor and Tong 2004) and only assess a minority of the entire brain signals (Logothetis 2008). Since both technologies are not without limitations, recent research has begun to integrate the two to enhance the validity of results (i.e., simultaneous fMRI-EEG; Huster et al. 2012).

Underlying Assumptions To understand the basic research process involving the use of neuroscientific technologies, it is important to distinguish two primary approaches in neuroscience research (Raichle 2010). The first and the most predominant approach is a reflexive view, in which researchers are interested in the brain regions activated in response to certain stimuli. In this case, the regions responsible for performing a task are commonly identified through a contrast between brain activities evoked by an experimental stimulus and brain activities at rest (i.e., simple subtraction method; Logothetis 2008). An alternative, yet a less common, approach is an intrinsic view, in which the main interest is in identifying the common organization of the brain in a resting state (Cole et al. 2014; Raichle 2010).

The distinction between the two approaches, reflexive and intrinsic, is analogous to the distinction between experiments and survey methods. That is, experiments elicit participants' responses to artificially created stimuli, and a reflexive view of the brain is based on observations of brain activities evoked by experimental stimuli. Accordingly, reflexive brain activity is measured while participants are performing one or more tasks. The intrinsic approach attempts to correlate individuals' general attitudes and behavioral tendencies with their ongoing brain activities, just as surveys are used to correlate individuals' trait, cognition, affect, and behavior. More specifically, intrinsic brain activities are measured while participants are completely relaxed, with their eyes closed or fixated to a still object for approximately 3–10 min. With that said, it is possible to combine or compare the two approaches in that brain activity in a resting state can serve as a reference point to be compared with evoked, reflexive neural signals (Raichle 2010).

Since the two approaches have different validity issues, the strengths and limitations of each approach can be evaluated separately. The validity of the reflexive approach has been demonstrated through the major advances in cognitive neuroscience, and its advantages and achievements have been well-documented in the literature (e.g., Lieberman 2007). As suggested above, the main assumption of the reflexive approach is that a simple subtraction of brain signals evoked by a task from the ongoing brain signals represents the brain regions responsible for that particular task (Logothetis 2008). However, several researchers have noted the limitations of this approach (Biswal et al. 2010; Braeutigam et al. 2017).

That is, the basic assumption of the reflexive approach may not be adequately met, since neuronal activities are more complex than are generally thought. Since brain activities are not linear in nature, they do not necessarily move in a single direction. Furthermore, because of the possible interaction between event-related activities and ongoing brain activities (Huang et al. 2017), it is often difficult to make valid conclusions from the simple subtraction technique.

Another assumption that can be challenged is that the activation and deactivation of a particular brain region imply an increase and decrease in excitatory signals in that region, respectively. However, brain activities are in fact a result of the summation of both excitatory and inhibitory signals (Logothetis 2008). Therefore, there is no clear evidence that the regions are activated primarily due to an increase in excitatory signals of those particular regions; rather, it might reflect a decrease in inhibitory signals. Likewise, even if the fMRI signal indicated little change in the brain activity, it might be a result of a significant increase in inhibitory signals accompanied by an equivalent increase in excitatory signals. In this case, researchers may fail to detect the actual changes in activities that might bear significant implications.

With the recognition of the above limitations of the reflexive view, the intrinsic view started to reveal its prominence over the recent years (Biswal et al. 2010; Braeutigam et al. 2017). Its utility has been shown within health and disease research (Raichle 2015) and in the domain of leadership research (e.g., Waldman et al. 2017a). Moreover, the fact that resting state brain activities take up 20% of human energy consumption, while changes in brain activities due to specific tasks account for only an additional 5% of the energy consumption at best, suggests the importance of a deeper understanding of ongoing or intrinsic brain activity (Clarke and Sokoloff 1999; Raichle and Mintun 2006).

However, the intrinsic view also bears its own limitations. For example, some researchers have argued that a resting state is not solely composed of stable individual differences, but may also involve state-dependent factors (Buckner et al. 2013). Therefore, researchers must take caution when drawing conclusions connecting intrinsic brain activities to trait-based differences.

4.3.2 Types of Ethical Dilemmas

Now that we briefly discussed methodological concerns, we now turn to more controversial ethical issues that lie in researchers' hands, including how to deal with methodological concerns. Although these issues might be entirely novel for most organizational scholars, they have been under serious scrutiny in a broader scientific and neuroethics community (e.g., Marcus 2002). Since organizational neuroscience has adopted the basic principles and methodologies from earlier neuroscience research, the ethical issues and principles largely share common ground. Furthermore, since the major ethical issues (e.g., human rights, privacy, and physical risks) are typical in many research streams that introduce new technologies (e.g., genetic studies), similar ethical principles can be derived from past experiences in relevant fields (Farah 2005).

Treatment of Participants One might argue that the use of neuroimaging or neurosensing technologies may impose a significant harm on research participants. However, just like the traditional survey or experimental methods, the research protocols using such technologies go through the approval process of Institutional Review Boards (IRBs) to ensure the protection of participants. Therefore, issues such as participants' privacy rights and potential risks, if any, are dealt with during the IRB process.

In terms of comfort and safety of participants, there is little if any discomfort or safety issues associated with EEG assessment. On the other hand, while safety is also of little concern with regard to fMRI, some participants may experience psychological discomfort from the space confinement that accompanies the procedure (Waldman et al. 2011). Accordingly, although fMRI is a noninvasive procedure, researchers should screen participants a priori to determine the extent to which prospective participants could negatively react to fMRI assessment procedures.

Data Storage and Analysis Neural data are often thought of as more sensitive than other types of data (e.g., survey), since they may reveal participants' unconscious thoughts and feelings. Although all the collected data, as in any research, are to be de-identified and aggregated for analyses, the mere thought that one's physiological data are stored somewhere may be disturbing to some individuals (Waldman et al. 2015). However, this issue is also addressed as part of the IRB process, by requiring participants to sign a written consent form. In addition, by explaining and assuring participants that no identifiable information will be kept, and that the collected data will only be used for research purposes, this issue can be further alleviated.

A more critical issue might arise in the data analysis procedures, particularly when scrutinizing individual neural data. Since neural data often involve a large number of variables (e.g., in the thousands for qEEG data), data analysis procedures often necessitate the probing of each individual's neural patterns. One possible scenario is that a researcher finds out an unusual neural pattern of a participant that might indicate a potential neurological disorder, i.e., incidental findings. In this case, what actions should the researcher take, especially if he or she is not completely knowledgeable or eligible for diagnosing health-related problems? In the United States, since organizational neuroscience scholars are not subject to the HIPAA (Health Insurance Portability and Accountability Act of 1996) Privacy Rule,[1] there are no legal obligations or restrictions to disclose such information to the individual without his or her request. Rather, it is the researcher's ethical decision to notify the participant of the potential problem and consult with a professional neuroscientist upon agreement. This was the case in a study by Waldman et al. (2011), in which the researchers contacted the participant to consult with a neuro-therapist:

> As a case example, one of our participants was a manager who reported anger management problems. Obviously, someone with such issues might be challenged when assuming a leadership role, as emotional equanimity is an important quality. It is hard to imagine that anyone in a leadership position who "flies off the handle" could be seen as inspirational. With the help of a neuro-therapist, we were able to pinpoint the root cause of the behavior, which as it turns out could be traced to a childhood baseball injury. This injury had affected a portion of the brain particularly relevant to an individual's emotional stability, especially

[1]The HIPAA Privacy Rule limits the entities that are subject to the rule as "health plans, health care clearinghouses, and to any health care provider who transmits health information in electronic form in connection with transactions for which the Secretary of HHS has adopted standards under HIPAA." For more information, see US Department of Health and Human Services. Summary of the HIPAA Privacy Rule: https://www.hhs.gov/sites/default/files/privacysummary.pdf

> with regard to anger issues. Based on knowledge gained through prior qEEG research dealing with the neurological basis of anger management problems (Fava 1997; Lubar, Congedo, and Askew 2003; Martens 2001), with a series of neurofeedback sessions, the individual was able to rearrange neuropathways in the affected area, create new pathways with healthy neighboring neurons, and largely correct the problem. Accordingly, he was able to set the stage to become a more effective leader. (p. 69)

An effective way to deal with these issues is to ask participants a priori, during the consent procedure, about their preference regarding whether they would like to get notified of serious health conditions that would provide them with strong net benefits if detected and known to them (Wolf et al. 2008). If they agreed to get informed of such conditions, researchers should be cautious before communicating the conditions, as it may cause anxiety on the part of the participants. They can consult with an expert if needed, as in the above case, but the participant's identity should remain anonymous, and the cost of consultation must be programmed into the research project (Wolf et al. 2008). However, researchers are not obliged to examine the conditions beyond the research purpose (Illes et al. 2006).

Interpretation While there is some skepticism about using neuroimaging and neurosensing technologies in research, it is also true that brain data are often seen as more accurate (Farah 2005; Racine et al. 2005). Indeed, an experimental study showed that neuroscientific information positively biases nonexpert judgments regarding the validity of the explanation of a particular phenomenon (Weisberg et al. 2008). However, since the two major approaches and technologies (i.e., fMRI and qEEG) are not without limitations, it is the researcher's role to understand and acknowledge the shortcomings and provide an accurate interpretation from complex data sets.

One way to address technological limitations is to combine data from multiple sources, which is actually a common practice in any research process that seeks to improve validity. For example, self-reported survey data are sometimes viewed as unreliable and invalid, since participants are inclined to select socially desirable responses, are subject to biases, and provide careless responses. Similarly, experiments often lack external validity, due to the artificial environment that is hardly generalizable to the real world. Therefore, studies combining the advantages of both field survey and experiment are often considered more reliable, thus producing more valid results. In the same vein, the recent trend in the field of neuroscience is to combine fMRI and EEG (i.e., simultaneous fMRI-EEG) or other technologies (e.g., PET, MEG). However, since it is difficult for ON scholars to use more than one of these technologies due to the prohibitive costs, an alternative approach can be to combine neural data with traditional survey data or other physiological data (e.g., eye movement, facial expression, heart rate, etc.).

As common in any new field, another important issue that may arise is potential misinterpretation of results. This can be attributed to the lack of experienced scholars who are knowledgeable in both organizational and neuroscience research. In fact, Racine et al. (2005) have cautioned about this issue, since negligence can lead to misinterpretation, especially in social sciences where micro-level neural activities are used to explain macro-level social phenomena. Furthermore, journal reviewers are likely to overlook mistakes or intentionally concealed information that may significantly alter the conclusions. Therefore, to enhance transparency and encourage replication of findings, it is crucial to educate organizational scholars on neuroscientific knowledge and data analytic procedures, perhaps through partnerships with more experienced neuroscientists (Waldman et al. 2017b). In addition, researchers may voluntarily share with other scholars aggregated data involving intrinsic brain activities as a way to encourage replication and knowledge expansion, as suggested by Fox and Raichle (2007).

4.4 Ethical Issues in Neuroleadership Practice

A more vexing issue with respect to the adoption of neuroscience in leadership pertains to its practical application. That is, the potential usage of neuroscientific technologies in employee selection and training may raise controversial legal and ethical issues. Recall our opening scenario in which you as a job applicant were asked to have your brain assessed. Although it is yet premature to discuss the feasibility of such an immediate application, the discussion of ethical issues is never untimely. Therefore, in the following section, we will discuss (1) common issues that may arise prior to the brain assessment, (2) ethical issues pertaining to employee selection and placement, and (3) issues pertaining to brain training.

4.4.1 Pre-assessment: Potential Coercion

Issues common to both selection and training involve transparency of the consent procedure, due to a potential for coercion of brain assessments (Chatterjee 2013). For instance, organizations might want to advise poor performers to go through brain enhancement sessions, despite the employees' psychological resistance (Farah 2005). Even if organizations do not force them to participate in such training, employees may feel an implicit pressure. Likewise, job applicants may feel reluctant to an optional brain assessment procedure, but may not be able to opt out, thinking that a refusal might disqualify them as a candidate. In addition, it is also possible, as in the above example of the leader with an anger management problem, that organizations detect individuals with potential brain diseases or abnormalities in the process of brain assessments during employment decisions or neurofeedback training sessions. In this case, they could be notified of the fact and directed to a neuro-therapist for consulting.

In order to prevent organizational coercion and embarrassment due to unexpected results, it would be necessary that companies convey detailed information and require informed consent to individuals prior to any procedures. For instance, job applicants and incumbents should be clearly assured that there are no associated disadvantages for non-participation. Participants should also be asked, prior to the assessment, to indicate whether they would prefer to get notified of any abnormalities in their brain activities.

4.4.2 Neuro-screening: Employee Selection and Placement

Imagine that neural data of a job candidate point out him or her to be potentially violent or that part of his or her brain was damaged through child abuse or other forms of mistreatment (Nemeroff 2004). In this case, should the employer ignore the information, or should the person be taken out of consideration because of the potential harm to the organization and its employees?

A major concern surrounding the neuroscientific application to employee selection and placement is the potential for the identification of sensitive, disqualifying data. Brain data are inherently different from traditional selection tests that assess individuals' ability and personality, in that brain data may reveal sensitive personal information that may not be uncovered through traditional methodologies. That is, individuals have little control over what is being measured, since neural trajectories may reveal one's static way of thinking, whether conscious or unconscious. Furthermore, unlike implicit or explicit tests, faking is impossible with brain data, even with accumulated experience (Steffens 2004). Thus, while uncovering potentially important information for employers, because of the sensitivity of such information, the use and storage of brain data may be more of a concern, at least as compared to other traditional selection methods.

Further, existing legal statutes may come into play. For example, the current legal provisions in the United States such as US Equal Employment Opportunity Commission (EEOC) and the Genetic Information Nondiscrimination Act (GINA) of 2008 prohibit employment discrimination based on health-related information, criminal records, or genetic information. In the same vein, an equivalent legislative act may need to be put into effect that prohibits discrimination based on neural data (Appel 2008). In addition, employees must be assured that the collected data will be used only for designated purposes—that is, only "job-related" information will be analyzed, used, and stored. Placing an emphasis on the job relevance of neural data will not only help keep the company from any legal or ethical disputes but can also increase applicants' job pursuit intentions and organizational attractiveness (Elkins and Phillips 2000; Hausknecht et al. 2004).

Finally, companies adopting neuroscientific technologies may neglect the importance of precise and accurate analysis. Without knowledgeable expertise in place (i.e., through either full-time staff or consultants), complex sets of brain data on applicants can be easily misinterpreted (Farah 2005). Therefore, we recommend that laws should stipulate a restriction on company uses of neural-based assessment devices and analytic programs to mandate that the purchasing company be equipped with sufficient expertise.

4.4.3 Neurofeedback: Leadership Training

Another issue can be raised with regard to neurofeedback interventions as a leadership development tool. Concerns have been raised as to (a) the rationale for using neurofeedback as a treatment for ineffective leaders, (b) reliability and safety issues, and (c) degradation of human effort. First, there are some concerns about implementing neurofeedback training on leaders, particularly because it may be used to target leaders with poor performance (e.g., Lindebaum 2016). However, it is quite common for organizations to maximize their training and development expenditures by identifying employees who might benefit the most from such activities (e.g., especially poor performers). Thus, while an argument could be made that neurofeedback should be made available to all individuals in leadership positions, it is equally plausible that such efforts target those who may be most in need of training.

Part of the problem when considering why and how neurofeedback might be used for leaders is that clinical applications of this technique have been traditionally implemented to correct behavioral maladies. That is, neurofeedback has been used to correct problems like anxiety and attention deficit disorders. It follows that if neurofeedback is used to develop leadership, it is because poor leadership is to be considered a malady or disorder. To the contrary, we argue that neurofeedback can also be considered as a tool to enhance peak performance. Thus, leaders who receive neurofeedback might do so simply to enhance their performance or effectiveness. For a more complete understanding of the possibilities that are posed by neurofeedback, see Stikic et al. (2015).

Despite our overall favorability toward neurofeedback, we partially concur with Lindebaum's (2016: 544) characterization that neurofeedback could inappropriately be used to address a deficiency of "a brain profile of an arbitrarily selected category of effective leadership." Our disagreement is that his term "arbitrarily" appears to be somewhat cavalier or straw man in nature in that we are unaware of researchers who would recommend neurofeedback targeting an aspect of leadership that is unrelated to trainees' effectiveness. On the other hand, it may be appropriate to base neurofeedback on a global index of effective leadership that simultaneously takes into account multiple aspects of leader effectiveness (e.g., transformational, servant, empowering, and so forth). This global index could then be used as a desirable end state to set the direction for the development of protocols of neurofeedback training.

Second, the reliability and safety of neurofeedback training, especially as applied to leadership, is not yet known. Unlike neuromodulation using psychotropic drugs or repetitive transcranial magnetic stimulation (rTMS; Chatterjee 2004), neurofeedback is a noninvasive learning procedure. Just as individuals learn new skills through traditional training in the expectation of a reward, during neurofeedback training, the brain learns how to alter its neuro-pathways to achieve a desirable outcome in a training protocol. The goal is to align the trainee's brain pathways with those that are known (i.e., through prior research) to be characteristic of a healthy or peak performance state (Gruzelier 2014). In other words, neurofeedback is a direct way of altering one's intrinsic brain that can be beneficial for particular tasks, or across contexts, depending on the goal of the training and the protocol that is used.

However, what is not as clearly understood is whether such alterations may have negative side effects in terms of also increasing the propensity of unwanted cognition or behavior. Thus, any attempt to use neurofeedback for leader development purposes should clearly monitor for the possibility of unwanted side effects.

Last, neurofeedback training may degrade the value of personal effort and striving (Farah 2005). Although neurofeedback is based on the assumption that individuals can consciously control their brain activities and thereby enhance performance, it is possible that over-reliance on neurofeedback might discourage learning through traditional development methods (e.g., experiential training, social experiences, and multi-source feedback). However, we are not advocating that neurofeedback sessions replace existing leadership development methods. Rather, as argued by Waldman et al. (2011), neurofeedback might be combined with traditional leader development methods so as to obtain a synergetic effect. That is, a combined volitional effort to change fundamental neural activity, while simultaneously acquiring experiential job-related skills in an applied setting, may be most effective in improving leader effectiveness.

4.5 Conclusion

Neuroleadership research has suggested the utility of neuroscientific methods in organizational research, particularly where faking behaviors are of concern due to social desirability and impression management (Waldman et al. 2017b). Furthermore, the efficacy of neurofeedback interventions among healthy individuals suggests their potential as a possible means to complement the limitations of traditional leadership training and development techniques. This promising future, however, accompanies ethical concerns pertaining to the potential misuse of neuroscience technologies and resulting data. We have argued that given existing IRB oversight, the potential for ethical abuses among basic researchers is relatively minimal. However, as we move forward in the future with practical applications (e.g., selection, placement, and training), organizations could face a host of ethical concerns involving neural assessment and use of data.

Nevertheless, these concerns should not halt the advancement of further research and careful consideration of applications in the evolving field of organizational neuroscience, given the efficacy of neuroleadership research and its practical implications. Furthermore, since the potential ethical concerns are not completely distinct to neuroscience-based applications, most of the issues can be dealt with by identifying ethical concerns in advance and following precedents in other fields (Farah 2005). Lastly, active discussions of ethical issues, such as those presented in this chapter, represent a positive signal that the field is moving forward in a healthy direction.

References

Appel JM. When the boss turns pusher: a proposal for employee protections in the age of cosmetic neurology. J Med Ethics. 2008;34(8):616–8.

Bagozzi RP, Verbeke WJ, Dietvorst RC, Belschak FD, van den Berg WE, Rietdijk WJ. Theory of mind and empathic explanations of Machiavellianism: a neuroscience perspective. J Manag. 2013;39(7):1760–98.

Balthazard PA, Waldman DA, Thatcher RW, Hannah ST. Differentiating transformational and non-transformational leaders on the basis of neurological imaging. Leadersh Q. 2012;23 (2):244–58.

Biswal BB, Mennes M, Zuo X-N, Gohel S, Kelly C, Smith SM, et al. Toward discovery science of human brain function. Proc Natl Acad Sci USA. 2010;107(10):4734–9.

Braeutigam S, Lee N, Senior C. A role for endogenous brain states in organizational research: moving toward a dynamic view of cognitive processes. Organ Res Methods. 2017. https://doi.org/10.1177/1094428117692104

Buckner RL, Krienen FM, Yeo BT. Opportunities and limitations of intrinsic functional connectivity MRI. Nat Neurosci. 2013;16(7):832–7.

Chatterjee A. Cosmetic neurology: the controversy over enhancing movement, mentation, and mood. Neurology. 2004;63(6):968–74.

Chatterjee A. Brain enhancement in healthy adults. In: Chatterjee A, Farah MJ, editors. Neuroethics in practice. Oxford: Oxford University Press; 2013. p. 3–15.

Clarke DD, Sokoloff L. Circulation and energy metabolism of the brain. In: Agranoff BW, Siegel GJ, editors. Basic neurochemistry. Molecular, cellular and medical aspects. 6th ed. New York: Lippincott-Raven; 1999. p. 637–70.

Cole MW, Bassett DS, Power JD, Braver TS, Petersen SE. Intrinsic and task-evoked network architectures of the human brain. Neuron. 2014;83(1):238–51.

DeYoung CG, Hirsh JB, Shane MS, Papademetris X, Rajeevan N, Gray JR. Testing predictions from personality neuroscience: brain structure and the big five. Psychol Sci. 2010;21(6):820–8.

Dobbins GH, Lane IM, Steiner DD. A note on the role of laboratory methodologies in applied behavioural research: don't throw out the baby with the bath water. J Organ Behav. 1988;9 (3):281–6.

Elkins TJ, Phillips JS. Job context, selection decision outcome, and the perceived fairness of selection tests: biodata as an illustrative case. J Appl Psychol. 2000;85(3):479–84.

Farah MJ. Neuroethics: the practical and the philosophical. Trends Cogn Sci. 2005;9(1):34–40.

Fox MD, Raichle ME. Spontaneous fluctuations in brain activity observed with functional magnetic resonance imaging. Nat Rev Neurosci. 2007;8(9):700–11.

Glimcher PW, Camerer CF, Fehr E, Poldrack RA. Introduction: a brief history of neuroeconomics. In: Glimcher PW, Camerer CF, Fehr E, Poldrack RA, editors. Neuroeconomics: decision making and the brain. London: Academic; 2009. p. 1–12.

Gruzelier JH. EEG-neurofeedback for optimising performance. I: A review of cognitive and affective outcome in healthy participants. Neurosci Biobehav Rev. 2014;44:124–41.

Hannah ST, Balthazard PA, Waldman DA, Jennings PL, Thatcher RW. The psychological and neurological bases of leader self-complexity and effects on adaptive decision-making. J Appl Psychol. 2013;98(3):393–411.

Hausknecht JP, Day DV, Thomas SC. Applicant reactions to selection procedures: an updated model and meta-analysis. Pers Psychol. 2004;57(3):639–83.

Highhouse S. Designing experiments that generalize. Organ Res Methods. 2009;12(3):554–66.

Huang Z, Zhang J, Longtin A, Dumont G, Duncan NW, Pokorny J, et al. Is there a nonadditive interaction between spontaneous and evoked activity? Phase-dependence and its relation to the temporal structure of scale-free brain activity. Cereb Cortex. 2017;27(2):1037–59.

Huster RJ, Debener S, Eichele T, Herrmann CS. Methods for simultaneous EEG-fMRI: an introductory review. J Neurosci. 2012;32(18):6053–60.

Illes J, Kirschen MP, Edwards E, Stanford LR, Bandettini P, Cho MK, Ford PJ, Glover GH, Kulynych J, Macklin R, Michael DB. Incidental findings in brain imaging research: what should happen when a researcher sees a potential health problem in a brain scan from a research subject? Science (New York, NY). 2006;311(5762):783–4.

Jack AI, Rochford KC, Friedman JP, Passarelli AM, Boyatzis RE. Pitfalls in organizational neuroscience: a critical review and suggestions for future research. Organ Res Methods. 2017. https://doi.org/10.1177/1094428117708857

Johnston SJ, Boehm SG, Healy D, Goebel R, Linden DE. Neurofeedback: a promising tool for the self-regulation of emotion networks. NeuroImage. 2010;49(1):1066–72.

Kim MJ, Whalen PJ. The structural integrity of an amygdala–prefrontal pathway predicts trait anxiety. J Neurosci. 2009;29(37):11614–8.

Lieberman MD. Social cognitive neuroscience: a review of core processes. Annu Rev Psychol. 2007;58:259–89.

Lindebaum D. Critical essay: building new management theories on sound data? The case of neuroscience. Hum Relat. 2016;69(3):537–50.

Lindebaum D, Raftopoulou E. What would John Stuart Mill say? A utilitarian perspective on contemporary neuroscience debates in leadership. J Bus Ethics. 2017;144(4):813–22.

Logothetis NK. What we can do and what we cannot do with fMRI. Nature. 2008;453 (7197):869–78.

Marcus D, editor. Neuroethics: mapping the field. Proceedings of the Dana Foundation Conference. Chicago: University of Chicago Press; 2002.

Moll J, Weingartner JH, Bado P, Basilio R, Sato JR, Melo BR, et al. Voluntary enhancement of neural signatures of affiliative emotion using fMRI neurofeedback. PLoS One. 2014;9(5): e97343.

Nemeroff CB. Neurobiological consequences of childhood trauma. J Clin Psychiatry. 2004;65 (Suppl1):18–28.

Racine E, Bar-Ilan O, Illes J. fMRI in the public eye. Nat Rev Neurosci. 2005;6(2):159–64.

Raichle ME. Two views of brain function. Trends Cogn Sci. 2010;14(4):180–90.

Raichle ME. The restless brain: how intrinsic activity organizes brain function. Philos Trans R Soc Lond B Biol Sci. 2015;370(1668):20140172.

Raichle ME, Mintun MA. Brain work and brain imaging. Annu Rev Neurosci. 2006;29:449–76.

Sebastian V. New directions in understanding the decision-making process: neuroeconomics and neuromarketing. Procedia Soc Behav Sci. 2014;127:758–62.

Singer T, Lamm C. The social neuroscience of empathy. Ann N Y Acad Sci. 2009;1156:81–96.

Steffens MC. Is the implicit association test immune to faking? Exp Psychol. 2004 Jun;51 (3):165–79.

Stikic M, Berka C, Korszen S. Neuroenhancement in tasks, roles, and occupations. In: Waldman DA, Balthazard PA, editors. Organizational neuroscience. London: Emerald; 2015. p. 169–86.

Thakor NV, Tong S. Advances in quantitative electroencephalogram analysis methods. Annu Rev Biomed Eng. 2004;6:453–95.

Verbeke WJ, Rietdijk WJ, van den Berg WE, Dietvorst RC, Worm L, Bagozzi RP. The making of the Machiavellian brain: a structural MRI analysis. J Neurosci Psychol Econ. 2011;4(4):205–16.

Waldman DA, Balthazard PA, Peterson SJ. Leadership and neuroscience: can we revolutionize the way that inspirational leaders are identified and developed? Acad Manag Perspect. 2011;25 (1):60–74.

Waldman DA, Balthazard PA, Peterson SJ. Conclusions and a look forward. In: Waldman DA, Balthazard PA, editors. Organizational neuroscience. London: Emerald; 2015. p. 295–306.

Waldman DA, Wang D, Hannah ST, Balthazard PA. A neurological and ideological perspective of ethical leadership. Acad Manag J. 2017a;60(4):1285–306.

Waldman DA, Ward M, Becker WJ. Neuroscience in organizational behavior. Annu Rev Organ Psych Organ Behav. 2017b;4:425–44.

Weisberg DS, Keil FC, Goodstein J, Rawson E, Gray JR. The seductive allure of neuroscience explanations. J Cogn Neurosci. 2008;20(3):470–7.

Wolf SM, Lawrenz FP, Nelson CA, Kahn JP, Cho MK, Clayton EW, Fletcher JG, Georgieff MK, Hammerschmidt D, Hudson K, Illes J. Managing incidental findings in human subjects research: analysis and recommendations. J Law Med Ethics. 2008;36(2):219–48.

Zotev V, Phillips R, Young KD, Drevets WC, Bodurka J. Prefrontal control of the amygdala during real-time fMRI neurofeedback training of emotion regulation. PLoS One. 2013;8(11):1–14.

'Murder They Said': A Content Analysis and Further Ethical Reflection on the Application of Neuroscience in Management

5

Dirk Lindebaum, Virginia L. Brown, and Ismael Al-Amoudi

Contents

Abstract

In this chapter, we offer a content analysis of top-tier management journals to examine the extent to which advocates of neuroscience in management pay heed to the ethical ramifications of their work. Based upon our analysis, we are able to robustly refute the claim by Butler and colleagues (Hum Relat 70:1171–1190, 2017) that Lindebaum's (Hum Relat 69(3):537–50, 2016) concerns about the lack of ethical concerns in the proliferation and application of neuroscientific ideas and measurements are basically much ado about nothing. By way of this content analysis, we advance the debate on the ethical ramifications of applying neuroscience in management by demonstrating (1) which ethical issues are recognised and (2) which ones are not. Doing so has the potential to open up new directions in studying the ethical and practical ramifications of neuroscience in and around workplaces.

D. Lindebaum (✉) · I. Al-Amoudi
Grenoble Ecole de Management, U. Grenoble Alpes ComUE, Grenoble, France
e-mail: mail@dirklindebaum.EU; http://www.dirklindebaum.EU

V. L. Brown
Cardiff Business School, Cardiff University, Cardiff, Wales

J. T. Martineau, E. Racine (eds.), *Organizational Neuroethics*, Advances in Neuroethics, https://doi.org/10.1007/978-3-030-27177-0_5

5.1 Introduction

In this chapter, we examine the extent to which advocates of neuroscience in management pay heed to the ethical ramifications of their work. By referring to ethical ramifications, we build on Roskies' (2002) two-branch distinction of neuroethics as the ethics of neuroscience and the neuroscience of ethics. However, our argument focuses on the first branch, being 'concerned with ethical, legal and social policy implications of neuroscience' (Illes and Bird 2006: 511).[1] While this branch also centres upon the ethical aspects of research practice (such as privacy and anonymity of data; see Hallinan et al. 2014), in this chapter, we are explicitly interested in the application of neuroscientific findings and technologies and the ethical issues they can create in the context of the workplace. This is a key boundary condition that needs to be taken into account as we present and discuss our analysis; although there is much to be said about both the neuroscience of ethics and the 'ethics of practice' (Roskies 2002), these are beyond the scope of this chapter.

An examination of the ethical implications of using neuroscience in management studies is justified as advocates do not always discuss such concerns (see, e.g. Waldman et al. 2016a) or go as far as to suggest that 'criticisms regarding the ethics of neuroscience research in management and organizations ... may be overstated' (Butler et al. 2017: 21). In an essay published in Human Relations, and subtitled 'the curious case of the straw man murder', Butler et al. (2017) consider it an 'ominous warning' when sceptics argue that 'there is little—if any—deeper conversation about the ethics of neuroscience as such among advocates of organizational neuroscience, especially if neuroscience is applied to emotional or cognitive enhancing in healthy individuals rather than clinical purposes' (Lindebaum 2016: 544). However, Butler et al. (2017) fail to appreciate the distinctiveness of 'organizational neuroscience' in that quote. While the burgeoning discussions about neuroethics among neuroethicists and neuroscientists more broadly speaking (cf. references lists in Lindebaum and Raftopoulou 2017; Lindebaum 2013) are to be welcomed, still of concern is the fact that advocates, keen to import neuroscientific ideas and methods into management, lag behind in their interest for ethical issues compared to neuroethicists (Lindebaum 2016; Lindebaum et al. 2018).

Butler et al. (2017) convey the impression that organisational neuroscientists are actively reflexive about the ethical, legal and social policy implications of their work by relying on two rhetorical devices. First, in their essay, they cite the (ethically reflexive) work of Roskies (2002), having consistently omitted it previously in their own major works in management (e.g. Senior et al. 2011; Lee et al. 2012a, b; Butler

[1]By contrast, the latter branch is concerned with the neurobiology of moral and ethical thinking and decision-making, how intuitions are generated and how individuals form judgements that specify which courses of action are prohibited, permissible or even obligatory (Akinci and Sadler-Smith 2012; Levy 2011). We construe this definition strictly in terms of 'practical ramification' of using neuroscience in management; we do not question the integrity of studies examined in terms of adherence to what is typically referred to a research ethics, such as issues around informed consent or fair treatment of participants (ESRC 2010).

et al. 2016). Second, they confuse the two above-mentioned branches of neuroethics identified by Roskies (2002). All studies cited by Butler et al. in their 2017 essay (i.e. Moll et al. 2005; Greene and Haidt 2002; Greene 2015) in defence of their claim that there is appreciation among organisational neuroscientists on the ethical ramifications of their work concern exclusively Roskies' (2002) second branch, namely, the neuroscience of ethics.[2] The attentive reader is, therefore, understandably puzzled by the fact that, in this instance, claims about the ethics of neuroscience are not supported by the sources they cite.

To advance the debate, we incorporate a content analysis in our chapter to examine the extent to which advocates of neuroscience in management pay heed to the ethical ramifications of their work. By way of this content analysis, we further advance the debate on the ethical ramifications of applying neuroscience in management by reflecting on (1) which ethical issues are actually recognised and (2) which ones are not. In particular, we identify the dual omission of human culture and of human reflexivity in neuroscientific approaches to management and organisation. By reflecting on the ethical implications of the omission of culture and reflexivity, we attempt to open up new directions in studying the ethical and practical ramifications of neuroscience in and around workplaces.

5.2 Content Analysis

Qualitative content analysis is defined 'as a research method for the subjective interpretation of the content of text data through the systematic classification process of coding and identifying themes or patterns' (Hsieh and Shannon 2005: 1278). Its stated aim is 'to provide knowledge and understanding of the phenomenon under study' (Downe-Wamboldt 1992: 314). Typically, studies employing qualitative content analysis centre upon the features of language as communication with particular heed to the content or contextual meaning of the text (see Hsieh and Shannon 2005 for a review). In this study, we follow Hsieh and Shannon's (2005) guidelines on so-called summative content analyses. Ordinarily, a summative qualitative content analysis first identifies and then quantifies specific words in a text for the purpose of distilling the contextual use of the words. Rather than inferring meaning, this approach seeks to explore usage of particular key words, for instance, when academic journals or textbooks are screened for key words (Rabow et al. 2000). Its advantage lies in it being an unobtrusive and impartial way to study the phenomenon of interest (Babbie 1992). There are several 'classical steps' involved in conducting a rigorous content analysis, such as (1) formulating the research questions to be answered, (2) choosing the sample to be examined, (3) defining the categories to be applied, (4) outlining the coding process and the coder training,

[2] We do recognise, however, that one of their articles discusses 'ethical' issues in the spirit we are concerned with here (Senior et al. 2011; see Table 5.1).

(5) implementing the coding process, (6) determining trustworthiness and finally (7) analysing the results of the coding process (Hsieh and Shannon 2005).

As indicated above, step (1) consists of examining the extent to which advocates of neuroscience in management pay heed to the ethical ramifications of their work. In the case of our content analysis, steps (4) and (5) were greatly simplified because we could perform keyword searches in our selected sample without 'coder training'. The rest of this section clarifies how we chose our sample (step 2) and keywords (step 3). We discuss how we ensured the trustworthiness (step 6) of findings under step iii below. The following section develops a discussion (step 7) of our results for organisational neuroethics.

Choosing the sample to be examined (step 2). In order to understand the extent of appreciation of neuroethics among advocates of neuroscience in management, we conducted an online search among all 20 so-called 4* journals in the UK-based Association of Business Schools' (ABS) journal list in March 2018. Academic journals are ranked according to their impact and quality and given a star rating from 1* to 4* (the highest). The ABS journal list assesses the quality of management publications worldwide and awards its 4* 'journal of distinction' accolade to those publications that uphold the highest scholarly standard and the widest diffusion. We have selected this list to tap into this fast-evolving research field as it unfolds within journals of highest international distinction and impact. We also wanted to ensure that we tap into journals for which the need for a substantive contribution is more distinct compared to journals or book chapters that do not belong into this category. We appreciate that including the latter categories would likely change the outcome of our analysis. At the same time, our focus can be justified as these journals operate at the frontiers of knowledge production, with many subsequent studies making references to research published in ABS-listed journals.

The ABS journal guide divides journals into different subject areas. The most relevant for our endeavour were (1) General Management, Ethics and Social Responsibility, (2) Organisation Studies, (3) Psychology (Organisational) and (4) Strategy. The inclusion of these five categories incorporates the journals in which the main thrust of the unfolding debate on the role and potential of neuroscience in management is subsumed. With this selection in place, we could determine the time frame for our investigation (starting with the first article detected—see Lane and Scott 2007—and ending in February 2018). Full articles (i.e. book reviews were excluded) were screened for whether 'neuroscience' provides (1) a key variable in a conceptual article, (2) the inspiration for articles blending a conceptual or methodological focus or (3) a method or source of data.

Following this first step, we found that 11 of the 20 eligible journals had published full articles concerned with neuroscience and management.[3]

[3]Those journals without any such articles were the *Academy of Management Review*, *Administrative Science Quarterly*, *British Journal of Management*, *Business Ethics Quarterly*, *Journal of Management Studies*, *Organization Studies*, *Journal of Occupational and Organizational Psychology*, *Journal of Occupational Health Psychology*, and *Journal of Vocational Behavior*.

The remaining journals and articles are featured in Table 5.1. The corresponding number of articles was 45 in toto at that stage. Articles mentioning the word 'neuroscience' in passing were not included, or where the context did not match our analysis (i.e. one article was devoted to the legal context).

Defining the categories to be applied and determining trustworthiness of findings (steps 3 and 6): We then subjected articles to another content analysis to distil whether the word 'ethic' (which flags up both 'ethics' and 'ethical') features in the respective article. We did this by using the 'search' function in the software we used to read the pdf files of articles. Where the term 'ethic' occurred, the article was perused to determine the context in which the term was employed (i.e. is it concerned with the appropriate branch of neuroethics?). To prevent articles slipping through the search net that displayed a concern for neuroethics, but which did not employ the term 'ethic', we then read all 45 articles in full to determine whether authors used a different term to express their appreciation of neuroethics as the ethics of neuroscience (and not the neuroscience of ethics).

Of these 45 articles, only 5 displayed appreciation of ethical concerns (and some only to a very minor extent) arising from applying neuroscientific theories and methods to inform organisational practice (see Table 5.1). Further four articles were classed as 'somewhat' discussing ethical concerns.[4] A more generous estimate would be that 9 of those 45 (i.e. 20%) articles display appreciation for neuroethics. However, 40 out of the 45 articles (or 88.9%) did not make any substantial mention of the ethical issue in the spirit with which we were primarily concerned (i.e. the ethics of applying neuroscience to management research and practice).[5] Thus, it is clear that advocates of neuroscience in management rarely attempt to provide a thoughtful account of how their work has the potential to adversely impact upon practice in the workplace, as evidenced by this neglect of any more thorough engagement with neuroethics, as defined above.

Identified and unidentified ethical issues in the analysis (step 7): Our content analysis indicates that ethical questions are rarely raised in management studies featuring neuroscientific data, methods or concepts. Even if these ethical issues are mentioned, in many cases, it amounts to little more than a fleeting reference. Despite this, our brief analysis is informative about how ethical questions are framed and, equally importantly, which issues are excluded from the conversation. Table 5.2 provides further details of the ethical issues identified by our analysis.

---

[4]Articles were read and coded according to an agreed approach, and we used 'somewhat' as a one-word term describing those articles which refer to ethics to a 'certain extent' and, in passing, as an after-thought at the very end of an article. Such articles would usually refer the reader to other sources, but certainly not a central element of their argument.

[5]Note that the Butler et al. article (Butler et al. 2017) complies with our search criteria. Hence, we include their article among the five. However, on reading the article, as mentioned already, the reader will see that they confuse Roskies' (2002) two branches of neuroethics (i.e. studies they cite touch on the neuroscience of ethics) and offer little relevant debate on neuroethics in the context of management.

Table 5.1 Overview of studies included in content analysis

Authors	Journal	Type of article	Neuroscientific focus	Discusses ethical concerns?[a]	Context
Laureiro-Martínez et al. (2015)	Organization Science	Conceptual article	Yes	No	n/a
Ocasio (2011)	Organization Science	Conceptual article	Yes (if only somewhat)	No	n/a
Senior et al. (2011)[b]	Organization Science	Conceptual article	Yes	Yes	This article identifies use of neuroscientific or genetic screening to detect so-called beneficial managerial traits (p. 809) as a significant implication for practice at work. This article recognises the existence of 'neuroethics' in the spirit we are concerned with here
Healey and Hodgkinson (2014)	Human Relations	Critical review	Yes	No	The word 'ethic' appears once in the abstract (without elaboration) and on two occasions in the reference list. Hence, we did not include it in our analysis
Lindebaum (2016)[b]	Human Relations	Critical essay	Yes	Yes	The essay discusses the ethical issue when unreliable data creates 'bad management practice'
Lindebaum and Zundel (2013)[b]	Human Relations	Conceptual article	Yes	Yes	This article calls, inter alia, upon researchers to reflect more critically upon what their attempts at optimizing and normalizing behaviour may do to those who become subjected to neuroscientific re-engineering treatments
Butler et al. (2017)[b]	Human Relations	Conceptual article	Yes	Yes	The article discusses 'ethical issues', suggesting that 'cognitive neuroscience' pays 'considerable attention' to them

Balthazard et al. (2012)	Leadership Quarterly	Empirical article (using EEG)[c]	Yes	No	n/a
Boyatzis et al. (2012)	Leadership Quarterly	Empirical article (using fMRI)[d]	Yes	No	n/a
Lee et al. (2012a)	Leadership Quarterly	Conceptual article	Yes	No	n/a
Waldman et al. (2011)	Leadership Quarterly	Conceptual article	Yes	No	n/a
Volk and Köhler (2012)	Organizational Research Methods	Conceptual article	Yes	No	n/a
Waldman et al. (2016b)[b]	Organizational Research Methods	Conceptual/ methodological article	Yes	Yes	This article suggests that organisation scholars using neuroscience need to attend to neuroethics. Areas of concerns identified are in relation to privacy and confidentiality, potential shifts in ethical norms to allow for benefit to others and obligations should neuro-assessment identify any neural pathology
Ahlfors and Mody (2016)	Organizational Research Methods	Conceptual/ methodological article	Yes	No	n/a
Christopoulos et al. (2016)	Organizational Research Methods	Conceptual/ methodological article	Yes	No	n/a
Konovalov and Krajbich (2016)	Organizational Research Methods	Conceptual/ methodological article	Yes	No	n/a
Massaro and Pecchia (2016)	Organizational Research Methods	Conceptual/ methodological article	Yes	No	n/a
Quaresima and Ferrari (2016)	Organizational Research Methods	Conceptual/ methodological article	Yes	No	n/a
Spence (2016)	Organizational Research Methods	Conceptual/ methodological article	Yes	No	n/a

(continued)

Table 5.1 (continued)

Authors	Journal	Type of article	Neuroscientific focus	Discusses ethical concerns?[a]	Context
Veniero et al. (2016)	Organizational Research Methods	Conceptual/ methodological article	Yes	No	n/a
Bagozzi and Lee (2017)	Organizational Research Methods	Conceptual/ methodological article	Yes	Somewhat	The authors refer to 'critical neuroscience' as working to address the implications of neuroscience, albeit not directly organisational neuroscience including, but not limited to, ethical issues
Jack et al. (2017)	Organizational Research Methods	Conceptual/ methodological article	Yes	Somewhat	The authors make brief reference to ethical issues arising from the use of some invasive or interventional neuroscience research methods
Braeutigam et al. (2017)	Organizational Research Methods	Conceptual/ methodological article	Yes	No	n/a
Dulebohn et al. (2016)	Journal of Applied Psychology	Empirical article (using fMRI)	Yes	No	n/a
Hannah et al. (2013)	Journal of Applied Psychology	Empirical article (using EEG)	Yes	No	n/a
Haesevoets et al. (2017)	Journal of Applied Psychology	Empirical article (using fMRI)	Yes	No	n/a
Ashkanasy et al. (2014)[b]	Journal of Organizational Behavior	Conceptual article	Yes	Somewhat	The authors note that 'organizational scholars have an important role to play in shaping the ethical and practical development of neuroscience technologies' (p. 916). How this might be put into practice, however, is a point where the article remains silent

Niven and Boorman (2016)	Journal of Organizational Behavior	Article providing methodological recommendation for fMRI	Yes	Somewhat, if not indirectly	The authors briefly mention that the ethics of using neuroscience to inform organisational interventions has been raised and suggest that this is discussed in detail in previous studies. Given this, we did not include this article in our analysis
Becker and Cropanzano (2010)	Journal of Organizational Behavior	Incubator piece	Yes	No	n/a
Lindebaum and Jordan (2014)	Journal of Organizational Behavior	Conceptual article	Yes	No	n/a
Lane and Scott (2007)	Organizational Behavior and Human Decision Processes	Conceptual article	Yes (if only somewhat)	No	n/a
Beugré (2009)	Organizational Behavior and Human Decision Processes	Conceptual article	Yes	No	Only makes mention of the neurocognitive model of ethics, hence being concerned with the second branch of Roskies's (2002) definition of neuroethics
Dulebohn et al. (2009)	Organizational Behavior and Human Decision Processes	Empirical article (using fMRI)	Yes	No	n/a
Mason et al. (2009)	Organizational Behavior and Human Decision Processes	Empirical article (using fMRI)	Yes	No	n/a
Morgeson et al. (2013)	Personnel Psychology	Conceptual article	Yes (if only somewhat)	No	n/a
Waldman et al. (2018)	Personnel Psychology	Empirical article	Yes	No	n/a
Bagozzi et al. (2013)	Journal of Management	Empirical article (using fMRI)	Yes	No	n/a

(continued)

Table 5.1 (continued)

Authors	Journal	Type of article	Neuroscientific focus	Discusses ethical concerns?[a]	Context
Becker et al. (2011)	Journal of Management	Conceptual article	Yes	No	n/a
Lee et al. (2012b)	Journal of Management	Commentary	Yes	No	n/a
Molenberghs et al. (2015)	Journal of Management	Empirical article (using fMRI)	Yes	No	n/a
Nofal et al. (2018)	Journal of Management	Review article	Yes	No	n/a
Powell (2011)	Strategic Management Journal	Conceptual article	Yes	No	n/a
Laureiro-Martínez et al. (2015)	Strategic Management Journal	Empirical article (using fMRI)	Yes	No[e]	n/a
Hodgkinson, Healey (2011)	Strategic Management Journal	Conceptual article	Yes[f]	No	n/a
Waldman et al. (2016a)	Academy of Management Journal	Empirical article (using EEG)	Yes	No	n/a

Note:
[a]We discuss here in the table only the branch of neuroethics concerned with practical/ethical implications of neuroscience at work, and not in the spirit of the neuroscience of ethical decision-making. In addition, as already stated in the introduction of our chapter, when we ask whether ethical issues are discussed in this table, we do not question the integrity of studies examined in terms of adherence to what is typically referred to a research ethics, such as issues around informed consent or fair treatment of participants
[b]These articles have been included in our content analysis
[c]EEG stands for Electroencephalography
[d]fMRI stands for functional magnetic resonance imaging
[e]Apart from a single mentioning of "Ethics Committee" in relation to the experimental procedure, there is no mentioning of the ethical issues we are concerned with here
[f]The article briefly mentions neuroscientific evidence in relation to emotional/affective and cognitive processes as they might relate to strategic management

Table 5.2 Identified ethical issues

Citation	Article title	Ethical issues identified
Articles which discuss ethical concerns		
Senior et al. (2011)	Organizational cognitive neuroscience	Implications of such things as screening for desirable traits, despite unlikely efficacy of so-doing Existence of 'neuroethics'
Lindebaum and Zundel (2013)	Not quite a revolution: Scrutinizing organizational neuroscience in leadership studies	Point to the implications of reductive assumptions underlying organisational neuroscience research and question its ability to identify and develop effective leaders
Waldman et al. (2016b)	The added value of neuroscience methods in organizational research	Ethical implications of reducing organisational behaviour to brain-based activity Issues arising from neuroethics: privacy and confidentiality; differing research/study ethics; potential for discovery of pathological condition (of study participants)
Lindebaum (2016)	Critical essay: Building new management theories on sound data? The case of neuroscience	Suggests that many of the ethical issues arise from methodological problems which in turn inform poor practice. Cautions that advocates of organisational neuroscience have insufficiently engaged with ethical issues
Butler et al. (2017)	Organizational cognitive neuroscience drives theoretical progress, or: The curious case of the straw man murder	Separate out their organisational cognitive neuroscience (OCN) from other approaches to organisational neuroscience Use primarily neuroscience of ethics sources to argue that ethics of neuroscience is well considered in the literature
Articles which 'somewhat' discuss ethical concerns		
Ashkanasy et al. (2014)	Neuroscience and organizational behavior: Avoiding both neuro-euphoria and neuro-phobia	Acknowledge role of organisational scholars in shaping ethics when it comes to neuroscience technologies
Niven and Boorman (2016)	Assumptions beyond the science: Encouraging cautious conclusions about functional magnetic resonance imaging research on organizational behavior	Primary focus on the technical challenges surrounding the use of fMRI in organisational research, while pointing the reader to the ethical debates elsewhere
Bagozzi and Lee (2017)	Philosophical foundations of neuroscience in organizational research: Functional and non-functional approaches	Point to critical neuroscience work as touching on neuroethics and the scope for critical management scholars to contribute to such efforts
Jack et al. (2017)	Pitfalls in organizational neuroscience: A critical review and suggestions for future research	Discuss neuroscience methods as applied to organisational research and highlight ethical issues with more invasive (PET, lesion) or manipulative approaches

As set out above, in the few welcome instances in which neuroethics has been explicitly considered by advocates of neuroscience in management, it is the researcher (rather than the participant) who is deemed to be in possession of the moral capacity to recognise the perils of neuroscientific screening of work-related 'desirable' behaviours (Senior et al. 2011). There is a risk—in cases when 'expertise' is combined with what is 'desirable' behaviour at work—that this renders it more difficult for participants to argue against neuroscientific screening or interventions at work. Researchers also have to delineate privacy and confidentiality issues, thereby establishing a research-centric approach to collecting neuroscientific data (Waldman et al. 2016b). It is researchers, too, who are assumed to be best placed to point to the invasiveness or potential for manipulation through some popular neuroscience techniques (Jack et al. 2017) and to determine the role that management scholars should play in 'shaping the ethical as well as practical development of neuroscience technologies' (Ashkanasy et al. 2014: 916). The commonality among these studies is that they mostly treat human participants as unreflexive cultural and judgemental dopes (Garfinkel 1984/1967) wherein 'understanding the brain processes behind observed attitudes and behaviors and their implications for predicting and modifying behaviors in the workplace' (Becker et al. 2011: 934) is privileged over and above humans as reflexive subjects engaging in moral and instrumental reflection in relation to their social, symbolic and material environments (Archer 2003, 2000).

Overall, the restrictive view of human beings reflected in organisational neuroscience is particularly well captured by Butler et al.'s (2017) rehearsal of the claim that:

> In this scientific age, to fully understand behaviour means that we must explain it as a complex interaction among genes, the brain, its biochemical state, the person's family upbringing, the way society has treated him or her, and the stimuli that impinge upon the person. (As originally cited in Senior et al. 2008: 55)

While the ontology they propose is commendably stratified, it is also impoverished. Butler et al. are cautious to avoid reducing human phenomena to a single ontological stratum, such as the brain. Unfortunately, the canvass they propose leaves no room for the human person, his or her private life and the internal conversations which he or she engages with varying degrees of success. Indeed, while biology, personal history and social conditions matter, they crucially underdetermine the subjective course and resulting outcomes of human reflection. No matter how well we know a human subject's biological constitution, personal history and networks of relations in which she or he engages, predicting the outcome of personal reflections is fraught with difficulty, except on trivial matters that require little reflection. While biological, historical and social conditions bear causal effects, the reflexive powers of human subjects are irreducible to the former for at least two reasons that have been extensively studied by leading authors of the critical realist current invoked by Butler et al. (2017) as a suitable meta-theoretical framework.

The first reason for refusing to reduce human reflexivity to its biological, historical and social conditions is that human reflexivity depends necessarily on the system of symbolic objects (which constitute together what Archer (1995) calls a cultural system) on which human subjects rely whenever they act reflexively. Indeed, thinking does not happen in a symbolic void. Instead, we would argue that cultural objects (including language) are as necessary as neurons for specific human activities. So, while we can only agree with the materialist claim that 'no neurons, no reflexivity', we should immediately add the culturally realist claim that 'no cultural system, no reflexivity'.

Yet, adding cultural systems to Butler et al.'s list of explanations for human behaviour will still be fundamentally insufficient. It is so, we propose, because while the aforementioned causes of human behaviour are plausible ones, none of them in isolation can fully account for reflexive human activity. The efficient cause of reflexive human activity is the human person her or himself, not her neurons or her genes, not her personal history or the society in which she is evolving, nor the cultural symbols she manipulates. As Archer (1995: 195) puts it:

> People, in turn, are capable of resisting, repudiating, suspending or circumventing structural or cultural tendencies, in ways which are unpredictable because of their creative powers as human beings.

To argue otherwise amounts to suppressing the human subject by reducing it to the mere result of biological, social or cultural forces. Unless human persons are recognised as the irreducible locus of human reflection, it is unclear how they can be recognised as ethical subjects or, conversely, as the bearer of moral rights and obligations.

This leads us to the 'problem of privacy', which provides our second reason for refusing to reduce human reflexive powers to their conditions, biological or otherwise. The problem of privacy is summarised by Archer (2003: 22) as follows: 'It consists in the fact that although certain mental states, like being excited, can be externally attributed to me by observing my shining eyes and general agitation, this is hardly possible for mental states which have no necessary external behavioural manifestations, that is, those which roughly fall under the heading of "deliberating"'.

The irreducibility of first-person stances and the inadequacy of third-person stances for knowing mental states are problematic for behavioural approaches (including behavioural ethics and works on implicit attitudes). But the irreducibility of first-person stances is also problematic for (neuroscientific) pretences to capture first-person mental states on the basis of third-person observations. As Vidal and Ortega put it, 'the set of shared beliefs of the *neuro* community would begin with the assertion that we are essentially our brains or (if the formulation were extreme) that "everything, absolutely everything, is in the brain"' (2017: 231).

By noticing and reflecting on the omission of culture and of personal reflexivity, our intention is to show that recognising these in neuroscientific studies opens up directions for researchers with a dual interest in organisational and neurological phenomena. For instance, future enquiries at the intersection of the neuroethics and organisation studies could reflect on whether cultural specificities have been fully taken into account, or whether they have been reduced, as we suggest, to presumably universal and non-linguistic nervous impulses. Studies of the colonisation of the cultural by the neurological may find support and inspiration in research projects such as those designated as 'critical neuroscience' (see, e.g., Slaby and Choudhury 2012). They may also find useful inspiration in works of history of science (such as Vidal and Ortega 2017), which examine neurological knowledge as a historically produced discourse but also in post-colonial critiques of science as an instrument of cultural domination that negates, vilifies or distorts indigenous world views (Bhabha 1994; Prakash 1999). Finally, organisational scholars interested in the displacement of the cultural by the neurological may find interest in recent works that criticise the hasty application of neuroscience to management and leadership practice (Lindebaum and Zundel 2013; Lindebaum et al. 2018). Without presuming the yet undiscovered findings of future studies, it might turn out that the scientific objectivity of which neuroscientific approaches pride themselves is also its biggest ethical flaw. This would be the case if it could be proved that neuroscience operates as a discourse that unduly ignores or reinterprets people's cultural categories in terms of neurological categories deemed to be more appropriate.

Another line of enquiry suggested by our analysis concerns the role of personhood in complex organisational settings. While the question of personhood is recognised among neuroscientists (see Fuchs 2006), we have grounds to believe that it has been silenced or marginalised as a result of unbridled excitement with organisational neuroscience. More specifically, if we are increasingly treated as 'neurological subjects' (Chan and Harris 2011), this can shift perceptions of personhood towards a more 'biologized' version thereof, reflecting the belief that we are no more than our brains (Lindebaum and Zundel 2013; Vidal and Ortega 2017).

Also, *not* discussed is the broader ethical tension between neuroscience as offering insights into cognitions and actions and neuroscience as 'nested within and influenced by the economic and political domains of society. These latter domains can be disconnected from—if not discordant with—considerations for employing science and technology in ways that best meet humanitarian needs' (Giordano and Benedikter 2012: 3). Currently organisational neuroscience makes no mention of any broader political or economic context despite an 'academic-industrial complex' dedicated to neuroscience and with fast-growing economic capital (Giordano and Benedikter 2012). More broadly, a 'funding effect' (Krimsky 2003) is evidence of an intertwining of commerce and academia, wherein the notion of 'disinterestedness' is on the wane, with a lack of objectivity, if not bias, a potential consequence of the practice of mixing commerce with science.

Further to this, the application of neuroscience in and around work generates significant ethical and practical concerns because of its potential to be used as a predictive tool; that is 'to find out characteristics or predispositions that may not only not be known to the individual, but in fact may never even manifest' (Hallinan et al. 2014: 65). One prominent example here is the neuroscientist James Fallon from the University of California-Irvine, who had a longstanding interest in studying 'criminal' brains, especially those of psychopaths.[6] Having learned from his mother that his family, going back generations, has a history of having brought forward violent

[6]Retrieved from http://www.npr.org/templates/story/story.php?storyId=127888976 on 22 June 2017.

individuals, he set out to examine the brains of his family members, including his own. Astonishingly, all his family members (i.e. wife, mother, siblings and children) had 'normal' scans, but in terms of his own brain scan, he said that 'I look just like one of those killers'. The practical ramifications of this possibility for the context of work (and beyond) are only too readily apparent; the exclusion of future employees or managers in selection processes because their brain profile suggests an emerging or future pathology (which has not actually materialised yet) and, as the came of James Fallon suggest, may never materialise. Likewise, in the presence of a brain scan suggesting (but not demonstrating) a pathology, there is also the possibility that employees or managers are denied access to career promotions and pay rises.

Finally, we posit that, to be effective, neuroethics needs to be proactive and involved as the field of organisational neuroscience develops and not an 'after-the-fact reflection or analysis' (Giordano and Benedikter 2012: 1). That implies that neuroethics should play a more prominent role in the conception and execution of research projects, for it would help limit excessive technological determinism. After all, not all research projects that are technologically feasible are ethically defensible.

5.3 Conclusion

Organisational neuroscience raises deep and complex ethical issues, and there have been debates in the field as to whether these issues are squarely addressed in the literature. The content analysis reported in this chapter refutes the claim that a great deal of evidence exists showing a keen interest in, and appreciation of, neuroethics (as defined earlier) among those scholars who import neuroscientific theories and ideas into management. Again, recall that Butler et al. (2017) actually do not cite relevant studies to counter Lindebaum's (2016) observation that there is hardly any appreciation of neuroethics among *management* scholars. In light of the 45 articles examined (and the fact that only five properly featured content on the topic), we can challenge the arguments presented by Butler and colleagues and expose their misinterpretation of the argument around ethics in neuroscience. Butler et al. (2017) purport that their cherished perspective (i.e. that of organisational cognitive neuroscience) will drive theoretical progress—time will tell. We wonder, however, what remains of ethical concerns in the development of 'strong' theoretical foundations if advocates maintain that 'without the appropriate application of cognitive neuroscience, organizational science will find it far more difficult to advance at the same rapid rate that it has over the past century' (Lee et al. 2012b: 922). Is speed of progress really the single-most crucial concern and motivation for management studies? In the context of neuroscience as applied to management, this has been called into question (Niven and Boorman 2016). In addition, as Davis (2015: 310) notes, 'if theoretical progress is assessed by the volume of articles published in a given year, then we are facing a tsunami of progress. On the other hand, if progress is judged by answering important questions about the world, then some scepticism is warranted'. We subscribe to the latter notion, for 'important questions about the world' will in future increasingly concern what it means to be human (as opposed to a transhuman or dehumanised existence—see Al-Amoudi and Morgan 2018; Harari 2016; Lindebaum 2015), how biotechnology (Froomkin 2016) affects social relations in and around work in future and what the consequences of a 'neurobiological complex' are (i.e. how the human brain becomes an object and target for governing humans—see Rose and Abi-Rached 2014). The issues we identified may matter less in the race for 'theoretical progress' to some, but for us, they serve as the incentive to articulate views that both challenge the received wisdom of a dominant movement of thought and showcase the necessity to re-humanise management studies.

References

Ahlfors SP, Mody M. Overview of MEG. Organ Res Methods. 2016:1–21.

Akinci C, Sadler-Smith E. Intuition in management research: a historical review. Int J Manag Rev. 2012;14(1):104–22.

Al-Amoudi I, Morgan J, editors. Realist responses to posthuman society. Ex Machina. New York: Routledge; 2018.

Archer MS. Realist social theory: the morphogenetic approach. Cambridge: Cambridge University Press; 1995.

Archer MS. Being human: the problem of agency. Cambridge: Cambridge University Press; 2000.

Archer MS. Structure, agency and the internal conversation. Cambridge: Cambridge University Press; 2003.

Ashkanasy NM, Becker WJ, Waldman DA. Neuroscience and organizational behavior: avoiding both neuro-euphoria and neuro-phobia. J Organ Behav. 2014;35(7):909–19.

Babbie E. The practice of social research. New York: Macmillan; 1992.

Bagozzi RP, Lee N. Philosophical foundations of neuroscience in organizational research: functional and nonfunctional approaches. Organ Res Methods. 2017; https://doi.org/10.1177/1094428117697042.

Bagozzi RP, Verbeke WJ, Dietvorst RC, Belschak FD, van den Berg WE, Rietdijk WJ. Theory of mind and empathic explanations of Machiavellianism a neuroscience perspective. J Manag. 2013;39(7):1760–98.

Balthazard PA, Waldman DA, Thatcher RW, Hannah ST. Differentiating transformational and non-transformational leaders on the basis of neurological imaging. Leadersh Q. 2012;23(2):244–58. https://doi.org/10.1016/j.leaqua.2011.08.002.

Becker WJ, Cropanzano R. Organizational neuroscience: the promise and prospects of an emerging discipline. J Organ Behav. 2010;31(7):1055–9.

Becker WJ, Cropanzano R, Sanfey AG. Organizational neuroscience: taking organizational theory inside the neural black box. J Manag. 2011;37(4):933–61.

Beugré CD. Exploring the neural basis of fairness: a model of neuro-organizational justice. Organ Behav Hum Decis Process. 2009;110(2):129–39.

Bhabha HK. The location of culture. London: Routledge; 1994.

Boyatzis RE, Passarelli AM, Koenig K, Lowe M, Mathew B, Stoller JK, et al. Examination of the neural substrates activated in memories of experiences with resonant and dissonant leaders. Leadersh Q. 2012;23(2):259–72.

Braeutigam S, Lee N, Senior C. A role for endogenous brain states in organizational research: moving toward a dynamic view of cognitive processes. Organ Res Methods. 2017.

Butler MJR, O'Broin HLR, Lee N, Senior C. How organizational cognitive neuroscience can deepen understanding of managerial decision-making: a review of the recent literature and future directions. Int J Manag Rev. 2016;18(4):542–59.

Butler M, Lee N, Senior C. Organizational cognitive neuroscience drives theoretical progress, or: the curious case of the straw man murder. Hum Relat. 2017;70:1171–90.

Button KS, Ioannidis JP, Mokrysz C, Nosek BA, Flint J, Robinson ES, et al. Power failure: why small sample size undermines the reliability of neuroscience. Nat Rev Neurosci. 2013;14 (5):365–76.

Chan S, Harris J. Neuroethics. In: Brain waves module. 1: Neuroscience, society and policy. London: Royal Society; 2011.

Christopoulos GI, Uy MA, Yap WJ. The body and the brain. Organ Res Methods. 2016; https://doi.org/10.1177/1094428116681073.

Davis GF. Celebrating organization theory: the after-party. J Manag Stud. 2015;52(2):309–19.

Downe-Wamboldt B. Content analysis: method, applications, and issues. Health Care Women Int. 1992;13(3):313–21.

Dulebohn JH, Conlon DE, Sarinopoulos I, Davison RB, McNamara G. The biological bases of unfairness: neuroimaging evidence for the distinctiveness of procedural and distributive justice. Organ Behav Hum Decis Process. 2009;110(2):140–51.

Dulebohn JH, Davison RB, Lee SA, Conlon DE, McNamara G, Sarinopoulos IC. Gender differences in justice evaluations: evidence from fMRI. J Appl Psychol. 2016;101(2):151–70.

ESRC. Framework for Research Ethics (FRE). 2010; http://www.esrc.ac.uk/_images/Framework_for_Research_Ethics_tcm8-4586.pdf. Accessed 31 May 2012.

Froomkin AM. Introduction. In: Calo R, Kerr I, editors. Robot law. Cheltenham: Edward Elgar; 2016. p. x–xxiii.

Fuchs T. Ethical issues in neuroscience. Curr Opin Psychiatry. 2006;19(6):600–7.

Garfinkel H. Studies in ethnomethodology. Cambridge: Polity Press; 1984/1967.

Giordano J, Benedikter R. An early-and necessary-flight of the owl of Minerva: neuroscience, neurotechnology, human socio-cultural boundaries, and the importance of neuroethics. J Evol Technol. 2012;22(1):14–25.

Greene JD. The rise of moral cognition. Cognition. 2015;135:39–42.

Greene J, Haidt J. How (and where) does moral judgment work? Trends Cogn Sci. 2002;6 (12):517–23.

Haesevoets T, De Cremer D, Van Hiel A, Van Overwalle F. Understanding the positive effect of financial compensation on trust after norm violations: evidence from fMRI in favor of forgiveness. J Appl Psychol. 2017;103(5):578–90.

Hallinan D, Friedewald M, Schütz P, de Hert P. Neurodata and neuroprivacy: data protection outdated? Surveill Soc. 2014;12(1):55.

Hannah ST, Waldman DA, Balthazard PA, Jennings PL, Thatcher RW. The psychological and neurological bases of leader self-complexity and effects on adaptive decision-making. J Appl Psychol. 2013;98(3):393–411.

Harari Y. Homo deus. A short history of tomorrow. London: Harvill Secker; 2016.

Healey MP, Hodgkinson GP. Rethinking the philosophical and theoretical foundations of organizational neuroscience: a critical realist alternative. Hum Relat. 2014;67(7):765–92.

Hodgkinson GP, Healey MP. Psychological foundations of dynamic capabilities: reflexion and reflection in strategic management. Strateg Manag J. 2011;32(13):1500–16.

Hsieh H-F, Shannon SE. Three approaches to qualitative content analysis. Qual Health Res. 2005;15(9):1277–88.

Illes J, Bird SJ. Neuroethics: a modern context for ethics in neuroscience. Trends Neurosci. 2006;29 (9):511–7.

Jack AI, Rochford KC, Friedman JP, Passarelli AM, Boyatzis RE. Pitfalls in organizational neuroscience: a critical review and suggestions for future research. Organ Res Methods. 2017.

Konovalov A, Krajbich I. Over a decade of neuroeconomics. Organ Res Methods. 2016;22 (1):1–26.

Krimsky S. A conflict of interest. New Sci. 2003;179(2410):21.

Lane VR, Scott SG. The neural network model of organizational identification. Organ Behav Hum Decis Process. 2007;104(2):175–92.

Laureiro-Martínez D, Brusoni S, Canessa N, Zollo M. Understanding the exploration–exploitation dilemma: an fMRI study of attention control and decision-making performance. Strateg Manag J. 2015;36(3):319–38.

Lee N, Senior C, Butler M. Leadership research and cognitive neuroscience: the state of this union. Leadersh Q. 2012a;23:213–8.

Lee N, Senior C, Butler MJR. The domain of organizational cognitive neuroscience. J Manag. 2012b;38(4):921–31.

Levy N. Neuroethics: a new way of doing ethics. AJOB Neurosci. 2011;2(2):3–9.

Lindebaum D. Pathologizing the healthy but ineffective: Some ethical reflections on using neuroscience in leadership research. J Manag Inq. 2013;22(3):295–305.

Lindebaum D. Book review: sapiens: a brief history of humankind (by Yuval Noah Harari). Manag Learn. 2015;46:636–8.

Lindebaum D. Critical essay: building new management theories on sound data? The case of neuroscience. Hum Relat. 2016;69(3):537–50.

Lindebaum D, Jordan PJ. A critique on neuroscientific methodologies in organizational behavior and management studies. J Organ Behav. 2014;35(7):898–908.

Lindebaum D, Raftopoulou E. What would John Stuart Mill say? A utilitarian perspective on contemporary neuroscience debates in leadership. J Bus Ethics. 2017;144(4):813–22.

Lindebaum D, Zundel M. Not quite a revolution: scrutinizing organizational neuroscience in leadership studies. Hum Relat. 2013;66(6):857–77.

Lindebaum D, Al-Amoudi I, Brown VL. Does leadership development need to care about neuroethics? Acad Manag Learn Edu. 2018;17:96–109.

Mason MF, Dyer R, Norton MI. Neural mechanisms of social influence. Organ Behav Hum Decis Process. 2009;110(2):152–9.

Massaro S, Pecchia L. Heart rate variability (HRV) analysis. Organ Res Methods. 2016;22(1):1–40.

Molenberghs P, Prochilo G, Steffens NK, Zacher H, Haslam SA. The neuroscience of inspirational leadership: the importance of collective-oriented language and shared group membership. J Manag. 2015;43(7):2168–94.

Moll J, Zahn R, de Oliveira-Souza R, Krueger F, Grafman J. The neural basis of human moral cognition. Nat Rev Neurosci. 2005;6(10):799–809.

Morgeson FP, Aguinis H, Waldman DA, Siegel DS. Extending corporate social responsibility research to the human resource management and organizational behavior domains: a look to the future. Pers Psychol. 2013;66(4):805–24.

Niven K, Boorman L. Assumptions beyond the science: encouraging cautious conclusions about functional magnetic resonance imaging research on organizational behavior. J Organ Behav. 2016;37:1150–77.

Nofal AM, Nicolaou N, Symeonidou N, Shane S. Biology and management: a review, critique, and research agenda. J Manag. 2018;44(1):7–31.

Ocasio W. Attention to attention. Organ Sci. 2011;22(5):1286–96.

Powell TC. Neurostrategy. Strateg Manag J. 2011;32(13):1484–99.

Prakash G. Another reason. Science and the imagination of modern India. Princeton: Princeton University Press; 1999.

Quaresima V, Ferrari M. Functional near-infrared spectroscopy (fNIRS) for assessing cerebral cortex function during human behavior in natural/social situations. Organ Res Methods. 2016.

Rabow MW, Hardie GE, Fair JM, McPhee SJ. End-of-life care content in 50 textbooks from multiple specialties. JAMA. 2000;283(6):771–8.

Rose N, Abi-Rached J. Governing through the brain: neuropolitics, neuroscience and subjectivity. Camb Anthropol. 2014;32(1):3.

Roskies A. Neuroethics for the new millennium. Neuron. 2002;35(1):21–3.

Senior C, Lee N, Butler M. The neuroethics of the social world of work. Am J Bioeth. 2008;8(1):54–5.

Senior C, Lee N, Butler M. Organizational cognitive neuroscience. Organ Sci. 2011;22(3):804–15.

Slaby J, Choudhury S. Proposal for a critical neuroscience. In: Critical neuroscience: A handbook of the social and cultural contexts of neuroscience. London: Wiley Blackwell. 2012. p. 27–51.
Spence C. Neuroscience-inspired design. Organ Res Methods. 2016:1–24.
Veniero D, Strüber D, Thut G, Herrmann CS. Noninvasive brain stimulation techniques can modulate cognitive processing. Organ Res Methods. 2016:1–32.
Vidal F, Ortega F. Being brains. Making the cerebral subject. New York: Fordham University Press; 2017.
Volk S, Köhler T. Brains and games. Organ Res Methods. 2012;15(4):522–52.
Waldman DA, Balthazard PA, Peterson SJ. Social cognitive neuroscience and leadership. Leadersh Q. 2011;22(6):1092–106.
Waldman D, Wang D, Hannah S, Balthazard P. A neurological and ideological perspective of ethical leadership. Acad Manag J. 2016a;60(4):1285–306. https://doi.org/10.5465/amj.2014.0644.
Waldman DA, Wang D, Fenters V. The added value of neuroscience methods in organizational research. Organ Res Methods. 2016b:1–27.
Waldman DA, Wang D, Hannah ST, Owens BP, Balthazard PA. Psychological and neurological predictors of abusive supervision. Pers Psychol. 2018;71(3):399–421. https://doi.org/10.1111/peps.12262.

Consumer Neuroscience: Recent Theoretical and Methodological Developments for Research and Practice Using a Cube Model

6

Marco Hubert and Mirja Hubert

Contents

Abstract

In marketing and consumer research, the application and integration of neuroscientific methods, insights, theories, and tools has increased steadily. In particular, the investigation of implicit and unconscious processes using neuroscientific methods to clarify decision-making processes holds great interest for both researchers and practitioners. Yet despite this interest, problems and limitations continue to hinder the emergence of this field of research. Which insights and analyses can, or cannot, be achieved with neuroscientific research methods in a consumer research context needs to be clarified. This review chapter outlines recent scientific and methodological developments to inform interested researchers and practitioners about the benefits and limitations, as well as ethical considerations, associated with integrating neuroscientific tools into consumer research.

M. Hubert (✉) · M. Hubert
Department of Management, MAPP Centre, Aarhus BSS, Aarhus University, Aarhus, Denmark
e-mail: mah@mgmt.au.dk

J. T. Martineau, E. Racine (eds.), *Organizational Neuroethics*, Advances in Neuroethics, https://doi.org/10.1007/978-3-030-27177-0_6

6.1 Introduction

An emergent research area integrates neuroscientific methods, insights, and theories into marketing research efforts. This so-called consumer neuroscience domain already has yielded some promising research findings, thus attracting the attention of researchers, practitioners, and the popular media (Plassmann et al. 2015). However, these developments also have sparked debate about whether the insights provide general and future value, what their limitations are, and which ethical issues arise from their use. We therefore describe the emergence of consumer neuroscience in more detail and then propose a cube model of consumer neuroscience as a systematic approach for structuring the impacts of this new field of research.

6.1.1 Understanding the Consumer

For every company, familiarity with its target market and knowledge about its customers are critical to entrepreneurial success and the value delivery process (Kotler 1996; Kotler and Keller 2006). The art of marketing management in turn requires a profound understanding of customers' needs and their complex decision-making processes (McCarthy 1960), especially when we consider the saturation of market structures, the increasing relevance of new digital channels, and the emergence of omni-channel experiences across different offline and online touchpoints during the customer journey (Verhoef et al. 2015). Notably, information processing, the relevance of emotions, and consumer decisions during the buying process are important topics for both marketing research and marketing management (Verhoef et al. 2015).

Because market success necessitates a broad understanding of consumer behavior, marketing research should turn to interdisciplinary approaches to help clarify consumers, their decision-making processes, and their customer journey (Kenning et al. 2012). Marketing and marketing research have long embraced such an inter- and transdisciplinary focus by implementing and integrating methods, insights, and theories from other research disciplines, including sociology, economy, and psychology (Howard and Sheth 1969; Kahneman and Tversky 1979; Tversky and Kahneman 1981).

6.1.2 Measuring Unconscious and Implicit Decision-Making Processes

A profound understanding of complex consumer decision-making processes requires consideration of unconscious and implicit influences on consumer behavior (Hubert 2010). Despite their importance for presenting a complete picture of decision making, these processes are difficult to investigate, observe, and measure (Camerer et al. 2005). Early attempts to detail consumers' reasoning relied on specific conceptual models, such as the stimulus-organism-response model (Howard and Sheth 1969). In these models, the organism is a "black box" that cannot provide insights (Engel et al. 1968; Howard and Sheth 1969; Nicosia 1966). Instead, the models try to explain consumer behavior using observable processes and theoretical constructs (e.g., attitudes, intentions). In contrast with economic performance indicators (e.g., growth rates, return on investments, sales), the theoretical constructs were neither observable nor objectively measurable (Kenning and Plassmann 2005).

To make them more observable, marketing and consumer researchers started to expand their methodological repertoire by integrating and combining neurophysiological approaches with traditional marketing research methods (e.g., focus groups, surveys, experiments), in pursuit of a more profound investigation of consumers' decision making. For example, some early marketing studies used electroencephalography (EEG) (Kroeber-Riel 1979; Krugman 1971). Methods such as electrodermal responses (EDR), eye tracking, and bio-signal measurements (e.g., heart rate) are widely accepted for studies of consumer behavior (Bagozzi 1991; Gröppel-Klein 2005; Poels and Dewitte 2006). Developments in brain-imaging technologies are extending the possibilities for understanding how neurophysiological aspects influence consumers. Already, various studies apply functional magnetic resonance imaging (fMRI) or functional near-infrared spectroscopy (fNIRS), which measure mostly unconscious and automatic decision-making processes (Ahlert and Hubert 2010; Hubert and Kenning 2008).

6.1.3 Consumer Neuroscience

These recent developments sparked a new research area, termed consumer neuroscience. This research field applies tools and theories from neuroscience to better understand decision making and cognate disciplines (Kenning et al. 2007b; Plassmann et al. 2015). Consumer neuroscience thus represents a sub-area of the higher-order research fields of decision neuroscience and neuroeconomics. Both consumer neuroscience and decision neuroscience "use a multidisciplinary and multimodal perspective to tackle its research questions" (Yoon et al. 2012: 484). Whereas decision neuroscience focuses on understanding decision making in general, consumer neuroscience is devoted to consumers' decision making explicitly (Kenning et al. 2007b; Plassmann et al. 2015; Yoon et al. 2012). Similarly, neuroeconomics is "the application of neuroscientific methods to analyze and understand economically relevant behavior" (Kenning and Plassmann 2005: 344), including risky behavior, processing of trust-assuring and trust-decreasing arguments, and social interactions (Baumgartner et al. 2008; Camerer et al. 2005; Delgado et al. 2005; De Martino et al. 2006; Dimoka 2010; Kenning and Plassmann 2005; King-Casas et al. 2005; Platt and Huettel 2008; Riedl et al. 2010a; Sanfey et al. 2003). Other research areas also have started to implement neuroscientific methods, such as finance (neuro-finance) (Mohr et al. 2010b), information systems (neuro-IS) (Dimoka and Davis 2008; Dimoka et al. 2012; Riedl et al. 2010a), and leadership (neuro-leadership), all of which have contributed to extend consumer neuroscience research.

Consumer neuroscience seeks to integrate neuroscientific methods, insights, and theories into marketing and consumer research, though the area of neuromarketing by definition targets business environments (Daugherty and Hoffman 2017; Lim 2017), because it "refers to practitioner and commercial interest in neurophysiological tools ... to conduct company-specific marketing research" (Plassmann et al. 2012: 19). Both approaches try to shed light on the human organism as a black box and thus understand unconscious and automatic processes better (Hubert and Kenning 2008; Plassmann et al. 2015).

The earliest consumer neuroscience investigations were exploratory and descriptive, focused on investigating the functional specializations of brain areas to research topics such as the effect of brands (Deppe et al. 2005b; Koenigs and Tranel 2008; McClure et al. 2004; Yoon et al. 2006), communication activities, and advertisements (Deppe et al. 2005a, 2007; Kenning et al. 2007c; Rossiter et al. 2001), processing of price information (Knutson et al. 2007; Plassmann et al. 2007, 2008c), and purchase decisions (Knutson et al. 2007). More recent studies also consider socio-demographic variables (Riedl et al. 2010b; Wager et al. 2003) and personality and impulsivity traits (Hubert et al. 2013). Across consumer neuroscience, decision neuroscience, and neuroeconomics, researchers seek to develop a unified theory of human decision making (Camerer et al. 2005; Glimcher 2002; Kenning and Plassmann 2005; Sanfey et al. 2006). However, a persistent challenge for this developing research field is determining when, how, and why studies should integrate neuroscientific methods, insights, and theories, according to their benefits and limitations. We propose a cube model of consumer neuroscience to address this challenge.

6.2 The Cube Model of Consumer Neuroscience

Even as a growing number of studies have established a firm foundation for consumer neuroscience research, the field continues to struggle for scientific recognition, due to uncertainty about its potential and outcomes (Achrol and Kotler 2012; Blakeslee 2004; Gul and Pesendorfer 2008; Plassmann et al. 2015; Yoon et al. 2012). Establishing consumer neuroscience as a research field and demonstrating its value for science and practice require both further research and replication but also a more systematic approach (Fumagalli 2010). We propose a three-dimensional cube model of consumer neuroscience to systemize this field, according to the potential, limitations, and ethical issues associated with integrating neuroscientific methods, insights, and theories with marketing and consumer research (Fig. 6.1). In addition to establishing scientific recognition of the field, this systematic approach can help companies find existing studies and research questions that can help them answer business-relevant questions. It is similar to a voxel in an fMRI analysis.

The first dimension, the *x*-axis, identifies relevant sub-fields for marketing management: consumer goods marketing [business-to-consumer (B2C) marketing of

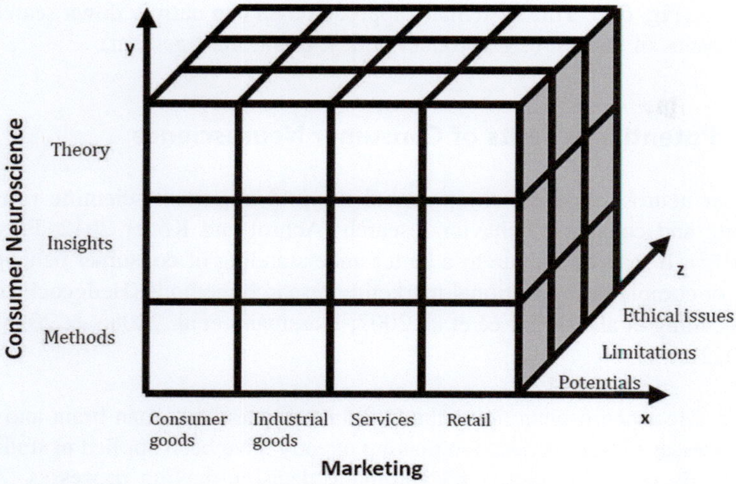

Fig. 6.1 Consumer neuroscience cube model: a systematic approach

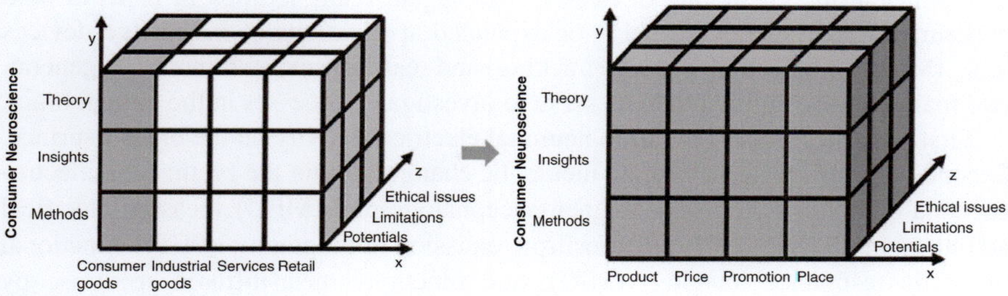

Fig. 6.2 Narrowing the search systematization of consumer goods marketing (left; grey area) to marketing mix instruments (right)

tangible goods], industrial marketing [business-to-business (B2B) marketing], service marketing (B2C or B2B), and retail marketing (Kotler 1996). The second dimension, the *y*-axis, integrates neuroscientific methods, insights, and theories that might be beneficial for investigating marketing- and consumer-related questions (Kenning et al. 2012). The third dimension, the *z*-axis, relates to potential benefits, limitations, and ethical issues that might emerge from this integration (Kenning et al. 2012).

The cube model of consumer neuroscience in turn can be adjusted flexibly to different research questions. For example, the *x*-axis might be adapted to investigate different components of a marketing sub-field, such as B2C marketing (Fig. 6.2, left). By identifying consumer goods marketing on the *x*-axis as an area of interest, the cube suggests new opportunities with regard to marketing mix instruments (product, price, communication, distribution; Kotler 1996), which establishes a new x-axis (Fig. 6.2). This systematic approach then can narrow down searches for specific issues of interest (e.g., product policy, brand management).

6.2.1 Potential Benefits of Consumer Neuroscience

Consumer neuroscience has the potential to yield important scientific results for marketing and consumer behavior research (Achrol and Kotler 2012; Plassmann et al. 2015). It may contribute to a better understanding of consumer behavior and support or complement traditional marketing research methods (Hedgcock and Rao 2009; Kenning et al. 2012; Lee et al. 2007; Plassmann et al. 2008a, b, 2015; Yoon et al. 2012).

Methods Most neuroscientific methods to investigate the human brain and underlying processes of human decision making already have been applied in studies that focus mainly on unconscious and automatic decision-making processes (Ambler et al. 2000; Ariely and Berns 2010; Hubert and Kenning 2008; Knutson et al. 2007; Kroeber-Riel 1979). Further developments could yield new research potential, including the use of mobile devices to investigate retail settings in realistic field experiments (Krampe et al. 2018) or expanded applications of interactive devices (e.g., fMRI in connection with eye tracking) and real-time measurements. In general, two main neuroscientific methods serve to investigate processes in the human brain.

First, some techniques measure neuronal electrical activity on the brain's surface. Second, other methods determine metabolic changes within the brain. Both electroencephalography (EEG) and magnetoencephalography (MEG) measure electrical activities in the human brain; positron emission tomography (PET), functional magnetic resonance imaging (fMRI), and functional near-infrared spectroscopy (fNIRS) measure metabolic changes (Huettel et al. 2009; Kenning et al. 2007a, 2012; Marchionni and Vromen 2010; Matthews and Jezzard 2004). All these approaches have advantages and disadvantages with regard to their temporality and spatial resolution. That is, EEG or MEG can measure the temporary intervals of signals, but imaging techniques like fMRI offer stronger spatial resolution, which is important for determining the activation of different subcortical regions. Another noninvasive method, repetitive transcranial magnet stimulation (rTMS), also can activate or deactivate specific brain regions with targeted stimulation. Using it in combination with fMRI, researchers can investigate observed decision processes and validate results that indicate the importance of specific brain regions. In this sense, rTMS might replace studies that investigate patients with specific brain lesions (Hallett 2000; Knoch et al. 2006; Walsh and Cowey 2000; Wassermann 1998).

In addition to technological progress, improvements to fMRI data analysis applications suggest promise for marketing and consumer research. For example, psycho- and physio-physiological interaction (PPI) (Friston et al. 1997; Gitelman et al. 2003), as well as dynamic causal modeling (DCM) (Stephan et al. 2005, 2008, 2010), go beyond identifying activated brain regions during decision processes. They support observations of temporal correlation and dynamic neural networks, so they can better depict the functional integration, or interconnectivity among different brain regions, that leads to human decisions (Banks et al. 2007; Hare et al. 2009, 2010).

Other advantages of the application of neuroscientific methods to investigate marketing-related questions, relative to more traditional methods, include, for example:

- A more objective measurement approach based on spatial and/or temporal localization of brain activity associated with consumer behavior (Ariely and Berns 2010; Hubert and Kenning 2008). In contrast, traditional measurements gather subjective perceptions from participants, without any direct measure of unconscious or automatic processes (Camerer et al. 2005; Lieberman 2007; Shiv and Fedorikhin 1999; Shiv et al. 2005).
- Simultaneous, real-time investigations of consumer decisions and associated neural processes. Participants then would not need to reconstruct the reasons, emotions, or feelings surrounding their decisions ex post (Hubert and Kenning 2008; Lee et al. 2007).
- Eliminating the risk of strategic behavior, because participants usually cannot consciously influence their brain activity. As a consequence, the results would be less biased by social desirability effects (Camerer et al. 2005).
- Improved theory validation and adaption. The possibility of replicating existing behavioral theories is compelling; widely publicized findings, such as the gender-specific differences that mark perceptions of trust and risk (Riedl et al. 2010b), could be confirmed or denied. By identifying the activation of non-hypothesized brain regions, exploratory consumer neuroscience studies also could produce new results that suggest new theoretical insights and enhancements to existing theories (Hubert 2010; Plassmann et al. 2015).
- Improved marketing research instruments. For businesses, neuroscientific methods establish a better trade-off between costs and utility. For example, measuring actual preferences or preference inconsistencies among consumers should improve the offerings of consumer goods and services (Ariely and Berns 2010; Plassmann et al. 2007).

Insights and Theories Insights and theories about the neural correlates of basic economic constructs like utility, preferences, risk, trust perceptions, social interactions, and value-based decisions are highly relevant for marketing research (Fehr and Camerer 2007; Fehr et al. 2005; Glimcher 2002; Glimcher et al. 2005, 2009; Huettel et al. 2006; Krain et al. 2006; Louie and Glimcher 2010; McCabe et al. 2001). For example, when people make purchase decisions under uncertainty, trust processes and social interactions likely have important roles (Dietvorst et al. 2009; Riedl et al. 2010b). These insights and theories also may be relevant for research in business management (Dimoka and Davis 2008), organizational studies, or finance (Knutson and Bossaerts 2007; Mohr et al. 2010a) that exhibit inter- and transdisciplinary connections (Dimoka et al. 2012). Figure 6.3 offers a graphic example of observed neural correlates and behavioral processes that already have been

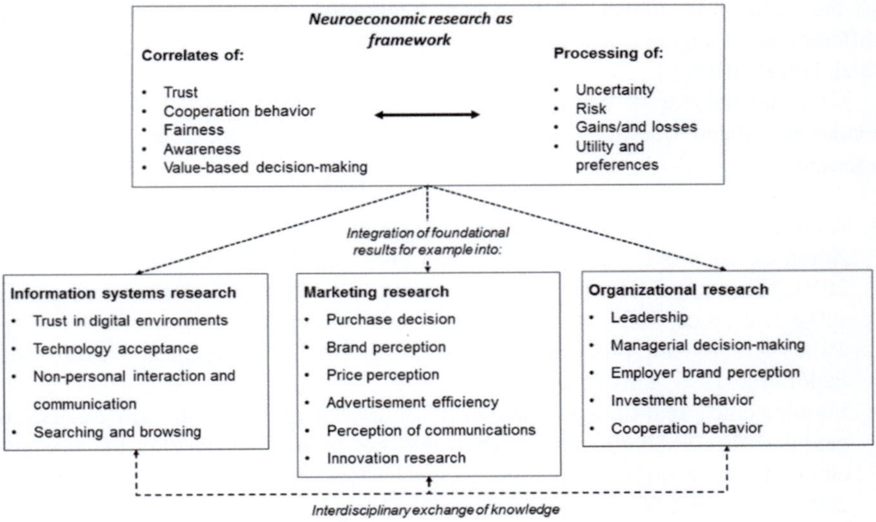

Fig. 6.3 Overview of observed neural correlates and processes

implemented and might add value to marketing research, as well as how the different research areas potentially interact.

Neuroscientific insights and theories about the antecedents of human decision making also might contribute to understanding of human behavior in the broader fields of decision neuroscience and neuroeconomics (Yoon et al. 2012). With regard to marketing research, these results need to be validated and adopted, especially in relation to the following theories:

- *Prediction error hypothesis*, which builds on assumptions about neural functionalities in dynamic learning processes and the interplay of expectations and experiences (Schultz et al. 1997). Dynamic learning processes lead to negative or positive prediction errors, which affect future behavior. For marketing research and practice, the prediction error hypothesis and reinforcement learning might clarify why consumers exhibit increased or diminished loyalty toward a brand or company.
- *Somatic marker hypothesis*, which states that somatosensory processes such as primary and secondary re-enforcers influence decision making (Bechara and Damasio 2005; Damasio 1996). It proposes that the integration of emotions and intuitions drives behavior. Research evidence affirms that emotions are highly relevant for rational economic decisions (Bechara et al. 1997, 2000). For marketing research and practice, this theory could help explicate the interplay of rational and emotional processes, such as at the point of sale (POS), and offer suggestions for how to "emotionalize" a brand.
- *Dual system theories*, grounded in impulsivity and addiction research, with the notion that two divergent systems (impulsive and reflective) interact to influence

decision-making processes (Bechara 2005; Camerer et al. 2005; Strack and Deutsch 2004; Strack et al. 2006). For marketing research, such theories could identify the motives for purchase behavior and its antecedents at the POS, as well as advance understanding of impulsive or compulsive buying behaviors.
- *Intertemporal choice theory*, pertaining to humans' perceptions of current and future outcomes (rewards, gains, losses) (Kable and Glimcher 2007). It tries to explain the value of future decisions today, using different discounting functions. Thus, intertemporal choice theory could detail the effects of sales campaigns, ambivalence in price perceptions, or sales negotiations for long-term contracts.

In particular, consumer neuroscience studies can apply neuroscientific insights and theories to understand and investigate marketing- and business-relevant questions with regard to purchase processes and different marketing mix instruments (Hubert and Kenning 2008; Kenning et al. 2012). For example, applying fMRI to investigate neural correlates of essential stages of the purchase process (product perception, price perception, decision) offers 30% improved predictions of the purchase decision, compared with traditional questionnaires (Grosenick et al. 2008).

For product and brand management, consumer neuroscience research can corroborate the impact of strong brands by observing the effects of improved salience or identification with the brand, product, or company (Ambler et al. 2004; Deppe et al. 2005b; Koenigs and Tranel 2008; McClure et al. 2004). Studies already have shown that only a first-choice brand can "emotionalize" decision making (Deppe et al. 2005b; Kenning et al. 2002). Only this favorite brand yields greater activity changes in regions associated with emotions and rewards, while simultaneously decreasing activity changes in regions associated with rational processes. This result challenges a general assumption of a linear effect of first-, second-, and third-choice brands, as is associated with the idea of an evoked or relevant choice set held in consumers' minds (Deppe et al. 2005b; Kenning et al. 2002). Other studies reveal that brands are processed as culturally based symbols (Schaefer et al. 2006; Schaefer and Rotte 2010) and are perceived differently when they are novel rather than familiar (Reimann et al. 2012). In addition, semantic judgments about products and persons are not processed similarly (Yoon et al. 2006). A recent study that combined fMRI and machine-learning techniques indicated that brand personality traits enter the human mind, suggesting an ability to predict people's thoughts about a brand on the basis of the brand associations and activation in their brains (Chen et al. 2015). The so-called marketing placebo effect (i.e., marketing-based expectations in the form of price and brand quality beliefs) also could be related to brain and behavioral associations; consumers with specific preconditions (i.e., high reward seeking, high need for cognition) appear more susceptible to marketing placebo effects (Plassmann and Weber 2015). Furthermore, perceptions of packaging designs could influence consumers' decision making (Hubert et al. 2013; Reimann et al. 2010; Stoll et al. 2008). Consumer neuroscience might inform the management of innovations as well (Ariely and Berns 2010), especially radical innovations, because it could support product testing and establish the optimal combination of brand and advertising campaigns. Continued studies that seek to improve product innovations and perceptions, communication effectiveness, and efficiency then could improve innovation management over time (Ariely and Berns 2010). Such studies offer interesting examples of how existing marketing theories can be validated or modified by integrating neuroscientific methods, insights, and theories with consumer and marketing research (Plassmann et al. 2015).

With regard to price management, consumer neuroscience research also has established the ambivalent effects of consumer price perceptions (Karmarkar et al. 2015). On the one hand, prices evoke negative perceptions and inhibit purchase decisions (Knutson et al. 2007). On the other hand, they can signal quality and positively influence product perceptions (Plassmann et al. 2008c). General neuroscientific insights and theories related to how people process rewarding stimuli and negative feelings help explain these contrary effects. With the use of fMRI and observed brain activation, these studies also enhance understanding of price limits (Linzmajer et al. 2011) and demonstrate that exposures to prices (i.e., price primacy) at different stages in the decision-making process have significant effects on price evaluations and product choices (Karmarkar et al. 2015).

Finally, for communication strategies and advertising campaigns, consumer neuroscience research promises to increase the effectiveness and efficiency of different tools (offline or online). Companies spend vast amounts on advertising, yet an estimated 26% of marketing budgets are allocated inefficiently and ineffectively (Ambler and Hollier 2004). Neuroscientific insights and methods might impose some initial costs, but ultimately they could specify the effects of various advertising campaigns in the brain (Deppe et al. 2005a, 2007; Kenning et al. 2007c). The most attractive advertisement campaigns would not only evoke positive evaluations but also prompt activity changes in brain regions associated with rewards. Unattractive stimuli instead induce activity changes in regions associated with processing negatively loaded stimuli, whereas neutral stimuli have no effect (Kenning et al. 2007c, 2009). A clear brand positioning in advertising campaigns thus is essential, to ensure retention by consumers. A recent study comparing different methods (self-reports, implicit measures, eye tracking, fMRI) indicates that fMRI offers the most value in terms of explaining advertising effectiveness, beyond a baseline level achieved by traditional methods (Venkatraman et al. 2015). In particular, activity in the ventral striatum is a strong predictor of market-level responses to advertising (Venkatraman et al. 2015).

6.2.2 Limitations of Consumer Neuroscience

The systematization of the consumer neuroscience field also requires consideration of its limitations (Fig. 6.1), especially with regard to methodological issues, including the potentially negative relation between costs and benefits, restrictions on technological capabilities, and experimental challenges (Kenning et al. 2007a). First, the costs of consumer neuroscience research, especially fMRI studies (Ariely and Berns 2010), remain very high. In some cases, the results, particularly for projects conducted in cooperation with business partners, may not justify the costs (Hubert and Kenning 2008).

Second, technological limitations arise from the trade-off between better spatial (fMRI) or better temporal (EEG, MEG) resolution. Many consumer neuroscience researchers prefer fMRI, which offers high spatial and decent temporal resolution (Huettel et al. 2009). However, when it comes to interpreting the data, fMRI relies on an indirect measure of brain activity, according to blood flow and oxygen use, such that it is subject to the blood oxygen level-dependent (BOLD) effect. Therefore, direct neural responses to the stimulus or experimental setup are not measurable, and the results often reflect aggregated analyses. This limitation must be taken into account during the data analysis and interpretation. In addition, studies often assume that measured effects can be traced back to the stimuli, without accounting for the potential influence of other, distracting, or noisy stimuli (Kenning and Plassmann 2005).

Third, in terms of experimental limitations, consumer neuroscience studies often suffer from small sample sizes, low external validity, and reliability problems, due to the lack of replication studies (Hubert 2010; Lim 2017; Plassmann et al. 2015). Some recent developments address some of these issues (Plassmann et al. 2015), but the studies remain limited, because participants do not make their decisions in natural environments. Unlike traditional behavioral experiments, fMRI-based investigations require participants to lie still; the scanner itself is loud and cramped. Such conditions could interfere with the data analysis (e.g., if the participant moves too much) and interpretation as well as challenge the generalizability of the results (Ariely and Berns 2010; Kenning et al. 2007a; Yoon et al. 2012). Another potential bias arises with regard to participant selection (e.g., no metallic appendages, cardiac pacemakers). The complex data analysis also demands substantial methodological competencies and expertise (Kenning and Plassmann 2005). Some experimental restrictions could be overcome with a multimethod approach that combines different neuroscientific (e.g., fMRI, EEG) and neurophysiological (e.g., eye tracking) methods. However, current research has not used multimethod approaches to validate results (Kenning and Linzmajer 2011; Kosslyn 1999; Logothetis 2008; Poldrack 2006; Savoy 2005).

Furthermore, limitations associated with integrating neuroscientific insights and theories into marketing and consumer research might reflect the more general critique about whether it is even possible to transfer existing insights and theories from one discipline to another (Gul and Pesendorfer 2008; Plassmann et al. 2012). For example, consumer neuroscience research seeks an "understanding of consumers' brains, not consumer behavior" (Plassmann et al. 2015: 429). Some arguments also suggest that there is no need to learn more about neural correlates of marketing-related processes because we can simply observe the relevant behavior (Plassmann et al. 2015).

Finally, some specific limitations reflect data acquisition and data interpretation approaches. For example, expectancy-based neural responses (Savoy 2005) arise if participants learn how they should react. If systematic, such learning behavior can appear in neural activity. If it is not intended by the researcher though, learning could complicate the data analysis. Ex post questionnaires might control for such learning mechanisms. Another issue is reverse or backward inference (Poldrack 2006), which reflects limitations for interpreting data. That is, activity changes observed in a new study design relate to specific cognitive functions that were predicted on the basis of existing studies. Using such reverse inferences implies the possible selection of only those results that fit the theory and the omission of other relevant results that might be counterintuitive or weaken the observed effect (Plassmann et al. 2015).

6.2.3 Consumer Neuroscience and Ethical Issues

Ethical issues associated with the integration of neuroscientific methods, insights, and theories into marketing research have been widely discussed (Ariely and Berns 2010; Farah 2002, 2007; Farah and Wolpe 2004; Kenning and Linzmajer 2011; Lim 2017; Murphy et al. 2008; Perrachione and Perrachione 2008; Racine et al. 2015; Shamoo 2010; Shamoo and Resnik 2009; Shamoo and Schwartz 2007; Wilson et al. 2008). The examination of these ethical issues is often referred to as neuroethics (Farah 2007; Illes and Racine 2005; Seixas and Bast 2008; Turner and Sahakian 2006). To understand neuroethics, we must consider the research frameworks in which the methods get applied. A more research-oriented, academic framework appears under the banner of consumer neuroscience (Hubert and Kenning 2008), whereas a more business-oriented framework involving third-party agencies (i.e., media agencies, consultancies) applies to neuromarketing (Hubert and Kenning 2008; Lim 2017; Ulman et al. 2015). Some general ethical issues pertain to neuroscientific methods, experiments, and the interpretation of the results, but neuromarketing prompts the majority of the discussion of ethical issues (Ulman et al. 2015).

In consumer neuroscience research, a stronger focus on emerging ethical issues largely is linked to experimental concerns and neuroscientific methods, because these technologies are primarily medical equipment, used to diagnose physiological or neurophysiological problems and diseases. They likely are associated with low risks, but the long-term effects of fMRI are not completely clear. Furthermore, the situation within the scanner (i.e., little space, noise) can be stressful for participants, which might undermine their well-being. If a scan reveals any abnormalities, participants would need to be informed, with the potential for vast negative consequences (Kenning and Linzmajer 2011). Other methods associated with greater risks, such as positron emission tomography (PET) and its invasive injection of a radiopharmaceutical, or rTMS with its influence on brain activity, rarely are used in marketing research (Kenning and Linzmajer 2011). Their application demands careful consideration, to balance the risks against the scientific value. In addition, all consumer neuroscience research should be monitored by an independent institution (e.g., research ethics boards, institutional review board) that confirms it meets high ethical research standards (Farah 2002, 2007; Farah and Wolpe 2004; Kenning and Linzmajer 2011; Murphy et al. 2008; Shamoo and Resnik 2009; Ulman et al. 2015).

When integrating neuroscientific insights and theories, the general assumption of a *homo neurobiologicus* who makes decisions based on biological determinants initiates divergent ethical opinions (Kenning and Plassmann 2005). Researchers try to predict behavior by analyzing neural activity patterns, brain stimulation, or connectivity among brain regions. Yet passionate defenders cite the importance of free will and consciousness and worry about the possibility of manipulating consumers (Bagozzi and Lee 2017; Olteanu 2015). Experimental setups that limit free will are scarce; this topic remains largely unexplored. Thus, the idea of a *homo neurobiologicus* whose social and economic behaviors solely result from neurobiology (Kenning and Plassmann 2005) is difficult to conceive. Furthermore, insufficient research addresses the complexity of the central nervous system and its interaction with other physiological systems (e.g., endocrine), as well as genetic aspects of development. Rigorous, intensive, transdisciplinary research, with larger samples and including replication studies, will help clarify the ethical issues involved in studying human decision-making processes on the basis of neurobiological insights. Yet considering the vast complexity of the human brain and other drivers of human behavior, we realize that there is still a long way to go to develop a unified theory of human decision making (Camerer et al. 2005; Glimcher 2002; Kenning and Plassmann 2005; Sanfey et al. 2006).

With a more business-oriented framework, neuroethics pertains more to privacy and data protection, as well as external ethical issues for interpreting the results. In particular, participants, their individual behaviors, and their personal data must be protected in all situations; no specific behavioral data may be linked to individual participants (Lim 2017; Olteanu 2015; Ulman et al. 2015). Business-related studies by researchers, media agencies, or consultancies must be monitored by independent institutions and held to high ethical research standards (Farah 2002, 2007; Farah and Wolpe 2004; Kenning and Linzmajer 2011; Murphy et al. 2008; Ulman et al. 2015), to ensure the autonomy, self-determination, privacy, confidentiality, and protection of any vulnerable groups of participants (Olteanu 2015; Ulman et al. 2015). Then, when it comes to the application and interpretation of the results, researchers must avoid oversimplification (Racine et al. 2015), such as promising an ability to find a "buy-button in the brain" (Blakeslee 2004) or read consumers' minds. Oversimplified explanations yield inaccurate perceptions and heat up the debate about the ethical applications of neuroscientific methods for marketing research (Ariely and Berns 2010). Furthermore, transparency, objectivity, and valid interpretations of the results limit suspicion of manipulation by companies, which is important for building trust (Olteanu 2015; Racine et al. 2015; Ulman et al. 2015). The fear that companies might manipulate people's opinions or read their thoughts is not justified by current neuroimaging techniques, which can only measure which brain regions are activated by a task. Still, companies must avoid creating any such fears, even as they continue to work to understand how neural activation actually leads to behavior (Hubert and Kenning 2008; Kenning and Plassmann 2005, 2012).

Even in the face of these ethical issues, consumer neuroscience offers the potential ability to resolve other ethical issues, especially in relation to consumer protection. For example, policy units (e.g., behavioral research units of national governments) might leverage information gathered through consumer neuroscience studies to promote greater consumer transparency. This domain also might help consumers understand their own decisions and behavior (Kenning and Plassmann 2005), potentially rendering them less susceptible to marketing efforts.

6.3 Conclusion and Further Research

A systematic approach, based on the cube model, can help structure and channel the potential, limitations, and ethical issues associated with the integration of neuroscientific methods, insights, and theories into consumer research. In particular, this systematic approach can help answer the following questions:

1. Which studies/insights already exist and can be integrated into marketing management?
2. When and how does the integration of neuroscientific methods, insights, and theories add value to investigations of marketing-relevant problems?
3. What limitations and ethical issues are associated with the integration of neuroscientific methods and theories?

In relation to technological developments, validation and replication of existing studies, and the integration and application of multimethod approaches, our proposed cube model offers a first step toward systemizing this research field. It can define research areas in which the integration of neuroscientific methods insights and theories will yield the most potential benefits; it also can reveal where limitations and ethical issues are most pressing and need to be considered before they can be applied to investigate marketing-relevant problems. Some primary challenges to consumer neuroscience research will be the effort to identify and explain the neural processes and mechanisms underlying consumers' decision making. Efforts to measure implicit processes, dissociate various psychological processes, understand individual differences, and improve predictions of human behavior continue to be limited (Plassmann et al. 2015). For consumer neuroscience research to contribute to understanding of consumer decision making, it must first become an established research sub-area within consumer and marketing research.

References

Achrol RS, Kotler P. Frontiers of the marketing paradigm in the third millennium. J Acad Mark Sci. 2012;40:35–52.

Ahlert D, Hubert M. Offene fragen der markenführung an die neurowissenschaftliche markenforschung. In: Bruhn M, Köhler R, editors. wie marken wirken – impulse aus der neuroökonomie für die markenführung. München: Franz vahlen; 2010. p. 47–55.

Ambler T, Hollier EA. The waste in advertising is the part that works. J Advert Res. 2004;44:375–89.

Ambler T, Ioannides A, Rose S. Brands on the brain: neuro-images of advertising. Bus Strateg Rev. 2000;11:17–30.

Ambler T, Bräutigam S, Stins J, Rose SP, Swithenby S. Salience and choice: neural correlates of shopping decisions. Psychol Mark. 2004;21:247–66.

Ariely D, Berns GS. Neuromarketing—the hope and hype of neuroimaging in business. Nat Neurosci. 2010;11:284–92.

Bagozzi RP. The role of psychophysiology in consumer research. In: Robertson TS, Kassarjian HH, editors. Handbook of consumer behavior. New Jersey: Prentice Hall; 1991. p. 124–61.

Bagozzi RP, Lee N. Philosophical foundations of neuroscience in organizational research: functional and nonfunctional approaches. Organ Res Methods. 2017:1–33.

Banks SJ, Eddy KT, Angstadt M, Nathan PJ, Phan KL. Amygdala-frontal connectivity during emotion regulation. Soc Cogn Affect Neurosci. 2007;2:303–12.

Baumgartner T, Heinrichs M, Vonlanthen A, Fischbacher U, Fehr E. Oxytocin shapes the neural circuitry of trust and trust adaptation in humans. Neuron. 2008;58:639–50.

Bechara A. Decision making, impulse control and loss of willpower to resist drugs: a neurocognitive perspective. Nat Neurosci. 2005;8:1458–63.

Bechara A, Damasio AR. The somatic marker hypothesis: a neural theory of economic decision. Games Econ Behav. 2005;52(2):336–72.

Bechara A, Damasio H, Tranel D, Damasio AR. Deciding advantageously before knowing the advantageous strategy. Science. 1997;275(5304):1293–5.

Bechara A, Damasio H, Damasio AR. Emotion, decision making and the orbitofrontal cortex. Cereb Cortex. 2000;10(3):295–307.

Blakeslee S. If you have a "buy button" in your brain, what pushes it? N Y Times. 2004;154:5.

Camerer C, Loewenstein G, Prelec D. Neuroeconomics: how neuroscience can inform economics. J Econ Lit. 2005;43(1):9–64.

Chen YP, Nelson LD, Hsu M. From "where" to "what": distributed representations of brand associations in the human brain. J Mark Res. 2015;52(4):453–66.

Damasio AR. The somatic marker hypothesis and the possible functions of the prefrontal cortex. Philos Trans R Soc Lond B Biol Sci. 1996;351(1346):1413–20.

Daugherty T, Hoffman E. Neuromarketing: understanding the application of neuroscientific methods within marketing research. In: Thomas AR, Pop NA, Iorga AM, Ducu C, editors. Ethics and neuromarketing. Cham: Springer; 2017. p. 5–30.

De Martino B, Kumaran D, Seymour B, Dolan RJ. Frames, biases, and rational decision-making in the human brain. Science. 2006;313(5787):684–7.

Delgado MR, Frank RH, Phelps EA. Perceptions of moral character modulate the neural systems of reward during the trust game. Nat Neurosci. 2005;8(11):1611–8.

Deppe M, Schwindt W, Krämer J, Kugel H, Plassmann H, Kenning P, Ringelstein EB. Evidence for a neural correlate of a framing effect: bias-specific activity in the ventromedial prefrontal cortex during credibility judgments. Brain Res Bull. 2005a;67(5):413–21.

Deppe M, Schwindt W, Kugel H, Plassmann H, Kenning P. Nonlinear responses within the medial prefrontal cortex reveal when specific implicit information influences economic decision making. J Neuroimaging. 2005b;15:171–82.

Deppe M, Schwindt W, Pieper A, Kugel H, Plassmann H, Kenning P, Deppe K, Ringelstein EB. Anterior cingulate reflects susceptibility to framing during attractiveness evaluation. NeuroReport. 2007;18(11):1119–23.

Dietvorst RC, Verbeke WJ, Bagozzi RP, Yoon C, Smits M, Van Der Lugt A. A sales force–specific theory-of-mind scale: tests of its validity by classical methods and functional magnetic resonance imaging. J Mark Res. 2009;46(5):653–68.

Dimoka A. What does the brain tell us about trust and distrust? Evidence from a functional neuroimaging study. MIS Q. 2010;34:373–96.

Dimoka A, Davis FD. Where does tam reside in the brain? The neural mechanisms underlying technology adoption. Proceedings of the 29th International Conference on Information Systems; 2008. p. 14–7.

Dimoka A, Banker R, Benbasat I, Davis F, Dennis A, Gefen D, Gupta A, Ischebeck A, Kenning P, Pavlou P, Müller-Putz G, Riedl R, vom Brocke J, Weber B. On the use of neurophysiological tools in is research: developing a research agenda for neurois. MIS Q. 2012;36(3):679–702.

Engel J, Kollat D, Blackwell R. Consumer behavior. New York: Holt, Rinehart & Winston; 1968.

Farah MJ. Emerging ethical issues in neuroscience. Nat Neurosci. 2002;5:1123–9.

Farah MJ. Social, legal, and ethical implications of cognitive neuroscience: "Neuroethics" for short. J Cogn Neurosci. 2007;19:363–4.

Farah MJ, Wolpe PR. Monitoring and manipulating brain function: new neuroscience technologies and their ethical implications. Hasting Cent Rep. 2004;34(3):35–45.

Fehr E, Camerer CF. Social neuroeconomics: the neural circuitry of social preferences. Trends Cogn Sci. 2007;11(10):419–27.

Fehr E, Fischbacher U, Kosfeld M. Neuroeconomic foundations of trust and social preferences: initial evidence. Am Econ Rev. 2005;95:346–51.

Friston KJ, Büchel C, Fink GR, Morris J, Rolls E, Dolan RJ. Psychophysiological and modulatory interactions in neuroimaging. NeuroImage. 1997;6(3):218–29.

Fumagalli R. The disunity of neuroeconomics: a methodological appraisal. J Econ Methodol. 2010;17(2):119–31.

Gitelman DR, Penny WD, Ashburner J, Friston KJ. Modeling regional and psychophysiologic interactions in fmri: the importance of hemodynamic deconvolution. NeuroImage. 2003;19 (1):200–7.

Glimcher PW. Review decisions, decisions, decisions: choosing a biological science of choice. Neuron. 2002;36:323–32.

Glimcher PW, Dorris MC, Bayer HM. Physiological utility theory and the neuroeconomics of choice. Games Econ Behav. 2005;52:213–56.

Glimcher PW, Camerer C, Poldrack RA, Fehr E. Neuroeconomics, decision making and the brain. 1st ed. London: Academic Press; 2009.

Gröppel-Klein A. Arousal and consumer in-store behavior. Brain Res Bull. 2005;67(5):428–37.

Grosenick L, Greer S, Knutson B. Interpretable classifiers for fMRI improve prediction of purchases. IEEE Trans Neural Syst Rehabil Eng. 2008;16(6):539–48.

Gul F, Pesendorfer W. The case for mindless economics. In: Caplin A, Schotter A, editors. The foundations of positive and normative economics. New York: Oxford University Press; 2008. p. 3–39.

Hallett M. Transcranial magnetic stimulation and the human brain. Nature. 2000;406:147–50.

Hare T, Camerer C, Rangel A. Self-control in decision-making involves modulation of the vmPFC valuation system. Science. 2009;324:646–8.

Hare TA, Camerer CF, Knoepfle DT, O'Doherty JP, Rangel A. Value computations in ventral medial prefrontal cortex during charitable decision making incorporate input from regions involved in social cognition. J Neurosci. 2010;30(2):583–90.

Hedgcock W, Rao AR. Trade-off aversion as an explanation for the attraction effect: a functional magnetic resonance study. J Mark Res. 2009;46(1):1–13.

Howard JA, Sheth JN. The theory of buyer behavior. New York: Wiley; 1969.

Hubert M. Does neuroeconomics give new impetus to economic and consumer research? J Econ Psychol. 2010;31(5):812–7.

Hubert M, Kenning P. A current overview of consumer neuroscience. J Consum Behav. 2008;7 (4/5):272–92.

Hubert M, Hubert M, Florack A, Linzmajer M, Kenning P. Neural correlates of impulsive buying tendencies during perception of product packaging. Psychol Mark. 2013;30(10):861–73.

Huettel SA, Stowe CJ, Gordon EM, Warner BT, Platt ML. Neural signatures of economic preferences for risk and ambiguity. Neuron. 2006;49(5):765–75.

Huettel SA, Song AW, McCarthy G. Functional magnetic resonance imaging. 2nd ed. Sunderland, MA: Sinauer Associates; 2009.

Illes J, Racine E. Neuroethics: dialogue on a continuum from tradition to innovation. Am J Bioeth. 2005;5(2):W3.

Kable JW, Glimcher PW. The neural correlates of subjective value during intertemporal choice. Nat Neurosci. 2007;10:1625–33.

Kahneman D, Tversky A. Prospect theory: an analysis of decision under risk. Econometria. 1979;47 (2):263–92.

Karmarkar UR, Shiv B, Knutson B. Cost conscious? The neural and behavioral impact of price primacy on decision making. J Mark Res. 2015;52(4):467–81.

Kenning P, Linzmajer M. Consumer neuroscience: an overview of an emerging discipline with implications for consumer policy. J Consum Prot Food Saf. 2011;6:111–25.

Kenning P, Plassmann H. Neuroeconomics: an overview from an economic perspective. Brain Res Bull. 2005;67(5):343–54.

Kenning P, Plassmann H, Deppe M, Kugel H, Schwindt W. The discovery of cortical relief. Field Res Neuromark. 2002;1:1–26.

Kenning P, Plassmann H, Ahlert D. Application of functional magnetic resonance imaging for market research. Qual Mark Res Int J. 2007a;10(2):135–52.

Kenning P, Plassmann H, Ahlert D. Consumer neuroscience. Marketing ZfP. 2007b;29(1):56–72.

Kenning P, Plassmann H, Kugel H, Schwindt W, Pieper A, Deppe M. Neural correlates of attractive ads. In: Koschnick WJ, editor. FOCUS-Jahrbuch 2007, Schwerpunkt: neuroökonomie, neuromarketing, neuromarktforschung. München: Focus Magazin Verlag; 2007c. p. 287–98.

Kenning P, Deppe M, Schwindt W, Kugel H, Plassmann H. The good, the bad and the forgotten: an fMRI study on ad liking and ad memory. Adv Consum Res. 2009;36:4–7.

Kenning P, Hubert M, Linzmajer M. Consumer neuroscience. In: Foxall G, editor. Handbook of new developments in consumer behavior. Edward Elgar: Cheltenham; 2012. p. 419–59.

King-Casas B, Tomlin D, Anen C, Camerer CF, Quartz SR, Montague PR. Getting to know you: reputation and trust in a two-person economic exchange. Science. 2005;308(5718):78–83.

Knoch D, Pascual-Leone A, Meyer K, Treyer V, Fehr E. Diminishing reciprocal fairness by disrupting the right prefrontal cortex. Science. 2006;314(5800):829–32.

Knutson B, Bossaerts P. Neural antecedents of financial decisions. J Neurosci. 2007;27 (31):8174–7.

Knutson B, Rick S, Wimmer GE, Prelec D, Loewenstein G. Neural predictors of purchases. Neuron. 2007;53(1):147–56.

Koenigs M, Tranel D. Prefrontal cortex damage abolishes brand-cued changes in cola preference. Soc Cogn Affect Neurosci. 2008;3(1):1–6.

Kosslyn SM. If neuroimaging is the answer, what is the question? Philos Trans R Soc Lond B Biol Sci. 1999;354(1387):1283–94.

Kotler P. Marketing management. Analysis, planning and control. Englewood Cliffs, NJ: Prentice-Hall; 1996.

Kotler P, Keller KL. Marketing management. Upper Saddle River, NJ: Prentice Hall; 2006.

Krain AL, Wilson AM, Arbuckle R, Castellanos FX, Milham MP. Distinct neural mechanisms of risk and ambiguity: a meta-analysis of decision making. NeuroImage. 2006;32(1):477–84.

Krampe C, Strelow E, Haas A, Kenning P. The application of mobile fNIRS to "shopper neuroscience"—first insights from a merchandising communication study. Eur J Mark. 2018;52:244–59.

Kroeber-Riel W. Activation research: psychobiological approaches in consumer research. J Consum Res. 1979;5(4):240–51.

Krugman HE. Brain waves measurement of media involvement. J Advert Res. 1971;11:3–9.

Lee N, Broderick AJ, Chamberlain L. What is "neuromarketing"? A discussion and agenda for future research. Int J Psychophysiol. 2007;63(2):199–204.

Lieberman MD. Social cognitive neuroscience: a review of core processes. Annu Rev Psychol. 2007;58:259–89.

Lim WM. Demystifying neuromarketing. J Bus Res. 2017;91:205–20.

Linzmajer M, Hubert M, Hubert M, Kenning P. The perception of lower and higher price thresholds: implications from consumer neuroscience. Adv Consum Res. 2011;39:792–3.

Logothetis NK. What we can do and what we cannot do with fMRI. Nature. 2008;453(12):869–78.

Louie K, Glimcher PW. Separating value from choice, delay discounting activity in the lateral interparietal area. J Neurosci. 2010;30:5498–507.
Marchionni C, Vromen J. Neuroeconomic: hype or hope? J Econ Methodol. 2010;17(2):103–6.
Matthews PM, Jezzard P. Functional magnetic resonance imaging. J Neurol Neurosurg Psychiatry. 2004;75(1):6–12.
McCabe K, Houser D, Ryan L, Smith V, Trouard T. A functional imaging study of cooperation in two-person reciprocal exchange. PNAS. 2001;98(20):11832–5.
McCarthy EJ. Basic marketing. Homewood, IL: Irwin; 1960.
McClure S, Li J, Tomlin D, Cypert KS, Montague LM, Montague PR. Neural correlates of behavioural preference for culturally familiar drinks. Neuron. 2004;44(2):379–87.
Mohr PNC, Biehle G, Heekeren HR. Neural processing of risk. J Neurosci. 2010a;30:6613–9.
Mohr PNC, Biehle G, Krugel LK, Li SC, Heekeren HR. Neural foundations of risk-return-trade-off in investment decisions. NeuroImage. 2010b;49:2556–63.
Murphy ER, Illes J, Reiner PB. Neuroethics of neuromarketing. J Consum Behav. 2008;7:293–302.
Nicosia F. Consumer decision processes. Englewood Cliffs, NJ: Prentice-Hall; 1966.
Olteanu MDB. Neuroethics and responsibility in conducting neuromarketing research. Neuroethics. 2015;8:191–202.
Perrachione TK, Perrachione JR. Brains and brands: developing mutually informative research in neuroscience and marketing. J Consum Behav. 2008;7:303–18.
Plassmann H, Weber B. Individual differences in marketing placebo effects: evidence from brain imaging and behavioral experiments. J Mark Res. 2015;52(4):493–510.
Plassmann H, Kenning P, Ahlert D. Why companies should make their customers happy: the neural correlates of customer loyalty. Adv Consum Res. 2007;34:1–5.
Plassmann H, Kenning P, Deppe M, Kugel H, Schwindt W. How choice ambiguity modulates activity in brain areas representing brand preference: evidence from consumer neuroscience. J Consum Behav. 2008a;7(4/5):360–7.
Plassmann H, O'Doherty J, Rangel A. Neural encoding of WTP computation during simple purchasing decisions. Adv Consum Res. 2008b;35:129.
Plassmann H, O'Doherty J, Shiv B, Rangel A. Marketing actions can modulate neural representations of experienced pleasantness. PNAS. 2008c;105(3):1050–4.
Plassmann H, Ramsøy TZ, Milosavljevic M. Branding the brain: a critical review and outlook. J Consum Psychol. 2012;22(1):18–36.
Plassmann H, Venkatraman V, Huettel S, Yoon C. Consumer neuroscience: applications, challenges, and possible solutions. J Mark Res. 2015;52(4):427–35.
Platt ML, Huettel SA. Risky business: the neuroeconomics of decision-making under uncertainty. Nat Neurosci. 2008;11:398–403.
Poels K, Dewitte S. How to capture the heart? Reviewing 20 years of emotion measurement in advertising. J Advert Res. 2006;46(1):18–37.
Poldrack RA. Can cognitive processes be inferred from neuroimaging data? Trends Cogn Sci. 2006;10(2):59–63.
Racine E, Bar-Ilan O, Illes J. fMRI in the public eye. Nat Rev Neurosci. 2015;6(2):159.
Reimann M, Zaichowsky J, Neuhaus C, Bender T, Weber B. Aesthetic package design: a behavioral, neural, and psychological investigation. J Consum Psychol. 2010;20(4):431–41.
Reimann M, Castaño R, Zaichkowsky J, Bechara A. Novel versus familiar brands: an analysis of neurophysiology, response latency, and choice. Mark Lett. 2012;23(3):745–59.
Riedl R, Banker R, Benbasat I, Davis F, Dennis A, Dimoka A, Gefen D, Gupta A, Ischebeck A, Kenning P, Pavlou P, Müller-Putz G, Straub D, vom Brocke J, Weber B. The application of neuroeconomics to IS-research. Commun Assoc Inf Syst. 2010a;27:243–64.
Riedl R, Hubert M, Kenning P. Are there neural gender differences in online trust? An fMRI study on the perceived trustworthiness of eBay offers. MIS Q. 2010b;34(2):397–428.
Rossiter JR, Silberstein RB, Harris PG, Nield G. Brain-imaging detection of visual scene encoding in long-term memory for TV-commercials. J Advert Res. 2001;41(2):13–22.

Sanfey AG, Rilling JK, Aronson JA, Nystrom LE, Cohen JD. The neural basis of economic decision-making in the ultimatum game. Science. 2003;300(5626):1755–8.
Sanfey AG, Loewenstein G, McClure SM, Cohen JD. Neuroeconomics: cross-currents in research on decision-making. Trends Cogn Sci. 2006;10:108–16.
Savoy RL. Experimental design in brain activation MRI: cautionary tales. Brain Res Bull. 2005;67 (5):361–7.
Schaefer M, Rotte M. Combining a semantic differential with fMRI to investigate brands as cultural symbols. Soc Cogn Affect Neurosci. 2010;5(2-3):274–81.
Schaefer M, Berens H, Heinze HJ, Rotte M. Neural correlates of culturally familiar brands of car manufacturers. NeuroImage. 2006;31(2):861–5.
Schultz W, Dayan P, Montague PR. A neural substrate of prediction and reward. Science. 1997;275 (5306):1593–9.
Seixas D, Bast MA. Ethics in fMRI studies. Clin Neuroradiol. 2008;18(2):79–87.
Shamoo AE. Ethical and regulatory challenges in psychophysiology and neuroscience-based technology for determining behavior. Account Res. 2010;17:8–29.
Shamoo AE, Resnik DB. Responsible conduct of research. 2nd ed. New York: Oxford University Press; 2009.
Shamoo AE, Schwartz J. Universal and uniform protections of human subjects in research. Am J Bioeth. 2007;7:7–9.
Shiv B, Fedorikhin A. Heart and mind in conflict: the interplay of affect and cognition in consumer decision making. J Consum Res. 1999;26:278–92.
Shiv B, Bechara A, Levin I, Alba JW, Bettman JR, Dube L, Isen A, Mellers B, Smidts A, Grant SJ, McGraw AP. Decision neuroscience. Mark Lett. 2005;16(3/4):375–86.
Stephan KE, Penny WD, Marshall JC, Fink GR, Friston KJ. Investigating the functional role of callosal connections with dynamic causal models. Ann N Y Acad Sci. 2005;1064:16–36.
Stephan KE, Kasper L, Harrison LM, Daunizeau J, den Ouden HE, Breakspear M, Friston KJ. Nonlinear dynamic causal models for fMRI. NeuroImage. 2008;42(2):649–62.
Stephan KE, Penny WD, Moran RJ, den Ouden HE, Daunizeau J, Friston KJ. Ten simple rules for dynamic causal modeling. NeuroImage. 2010;49(4):3099–109.
Stoll M, Baecke S, Kenning P. What they see is what they get? An fMRI-study on neural correlates of attractive packaging. J Consum Behav. 2008;7(4-5):342–59.
Strack F, Deutsch R. Reflective and impulsive determinants of social behavior. Personal Soc Psychol Rev. 2004;8(3):220–47.
Strack F, Werth L, Deutsch R. Reflective and impulsive determinants of consumer behavior. J Consum Psychol. 2006;16(3):205–16.
Turner DC, Sahakian BJ. Ethical questions in functional neuroimaging and cognitive enhancement. Poiesis Prax. 2006;4(2):81–94.
Tversky A, Kahneman D. The framing of decisions and the psychology of choice. Science. 1981;211(4481):453–8.
Ulman YI, Cakar T, Yildiz G. Ethical issues in neuromarketing: "I consume, therefore I am!". Sci Eng Ethics. 2015;21(5):1271–84.
Venkatraman V, Dimoka A, Pavlou PA, Vo K, Hampton W, Bollinger B, Hershfield HE, Ishihara M, Winer RS. Predicting advertising success beyond traditional measures: new insights from neurophysiological methods and market response modeling. J Mark Res. 2015;52 (4):436–52.
Verhoef PC, Kannan PK, Inman JJ. From multi-channel retailing to omni-channel retailing: introduction to the special issue on multi-channel retailing. J Retail. 2015;91(2):174–81.
Wager TD, Phan KL, Liberzon I, Taylor SF. Valence, gender, and lateralization of functional brain anatomy in emotion: a meta-analysis of findings from neuroimaging. NeuroImage. 2003;19 (3):513–31.
Walsh V, Cowey A. Transcranial magnetic stimulation and cognitive neuroscience. Nat Rev Neurosci. 2000;1:73–9.

Wassermann EM. Risk and safety of transcranial magnetic stimulation: report and suggested guidelines from the international workshop on the safety of repetitive transcranial magnetic stimulation. Electroencephalogr Clin Neurophysiol. 1998;108:1–16.

Wilson RM, Gaines J, Hill RP. Neuromarketing and consumer free will. J Consum Aff. 2008;42:389–410.

Yoon C, Gutchess AH, Feinberg F, Polk TA. A functional magnetic resonance imaging study of neural dissociations between brand and person judgements. J Consum Res. 2006;33:31–40.

Yoon C, Gonzalez R, Bechara A, Berns GS, Dagher AA, Dubé L, Huettel SA, Kable JW, Liberzon I, Plassman H, Smidts A, Spence C. Decision neuroscience and consumer decision making. Mark Lett. 2012;23(2):473–85.

Neuroenhancement at Work: Addressing the Ethical, Legal, and Social Implications

7

Veljko Dubljević, Iris Coates McCall, and Judy Illes

Contents

Abstract

Neuroenhancement is associated with a wide range of existing, emerging, and future biomedical technologies that are intended to improve human cognitive performance and mitigate—if not reverse—human error. Neuroenhancement in classrooms, universities, and the military has been discussed at length, but the workplace has been largely omitted from the conversation until now. By providing examples from branches of the commercial market that are rarely linked with cognitive enhancement in the literature, we argue that neuroenhancement at work is likely to become a major challenge in the labor market. Therefore, we focus here on the specific application of neuroenhancements to the workplace. Central issues involve both drugs and

V. Dubljević (✉)
Department of Philosophy and Religious Studies, North Carolina State University, Raleigh, NC, USA
e-mail: veljko_dubljevic@ncsu.edu

I. C. McCall · J. Illes
Neuroethics Canada, Division of Neurology, Department of Medicine, University of British Columbia, Vancouver, BC, Canada

J. T. Martineau, E. Racine (eds.), *Organizational Neuroethics*, Advances in Neuroethics, https://doi.org/10.1007/978-3-030-27177-0_7

devices, some of which are well-trodden ethical concerns while others are novel challenges. We conclude with a brief discussion and outline of a discourage-use policy that has the potential to mitigate the challenges of neuroenhancement at work.

7.1 Introduction

The term neuroenhancement has captured the imagination of a wide range of professionals—academics, journalists, and legal scholars—as well as the general public and others in this hyper-competitive world. If people are to succeed, they have to perform better, respond faster, and be smarter than everyone else. Neuroenhancement is associated with a wide range of existing, emerging, and future biomedical technologies that are intended to improve human cognitive performance and mitigate—if not reverse—human error (STOA 2009). Psychoactive substances are the most commonly discussed means of boosting attention, memory, and perseverance (Dubljević 2013a), whereas neurotechnologies that may interface with computers or other technological means of augmenting cognition (Enriquez-Geppert et al. 2017; Dubljević 2015a) are among the newest approaches.

Neuroenhancement in classrooms, universities, and the military has been discussed at length (BMA 2007; STOA 2009), but the workplace has been largely omitted from the conversation until now. In this chapter, we discuss the application of neuroenhancements to the labor force. Central issues here involve both drugs and devices, some of which are well-trodden ethical concerns and others novel challenges.

For the purposes of this chapter, we adopt the broadest possible definition of neuroenhancement that does justice to its nuances in both science and society. We use the term neuroenhancement to refer to any objective or subjective non-health-related improvement in cognitive functioning, mental acuity, increased wakefulness, or perseverance in task performance resulting from any drug, device, or neuroscientific procedure that exerts its function either directly or indirectly through neural activity and is undertaken to gain competitive advantage.

This definition has the virtue of dissociating contexts that are socially encouraged from those that are currently discouraged or even prohibited. Preventive, curative, rehabilitative, and compensatory uses of pharmaceutical drugs and devices are important elements in meeting health needs, whereas the intention behind neuroenhancement is not to meet health needs but to gain competitive advantage. Thus, the distinction between treatment and enhancement is not so much about the nature of the drug, technology, or intervention itself as it is about the purpose for which it is used.

The mixture of individual choice and social or organizational needs that may be driving the phenomenon of neuroenhancement has recently been explored by contemporary expert panels offering advice to decision makers (BMA 2007; STOA 2009; Academy of Medical Sciences 2012). Indeed, recent literature has provided numerous substantive claims about the social impact of neuroenhancement in terms of coercion and justice. By providing examples from branches of the commercial market that are rarely linked with cognitive enhancement in the literature, we argue that neuroenhancement at work is likely to become a major challenge (Appel 2008; Dubljević 2012a). We discuss the challenges related to the highest potential to the use of enhancements in the workplace here. Then, we discuss relevant policy options that pertain to relevant organizational goals and individual choices in liberal democratic societies. We review a recent policy proposal that favors the economic model of taxation (Dubljević 2012a, b) and challenge it with substantive objections based on both normative assumptions and practical consequences that such a policy might engender. Finally, we canvas take-home messages about neuroenhancement in the workplace that society should take into consideration if substantive policy disagreements persist.

7.2 Existing Neuroenhancers Relevant to the Workplace

7.2.1 Pharmaceutical Enhancement

Proponents of neuroenhancement claim that amphetamine, methylphenidate, and modafinil—the stimulant drugs most commonly used for nonmedical purposes—are effective enhancers of cognitive functioning. Expectations regarding the effectiveness of these cognitive enhancers, however, may exceed the actual effects produced. For instance, a recent study (Illieva et al. 2013) of 46 fully-rested Caucasian young healthy adults did not find any reliable effects of stimulants on a range of cognitive improvements, even though they reported perceived improvements. Other recent studies have shown that, in healthy individuals, these drugs do not actually promote the acquisition of new information and may even impair performance on tasks requiring adaption, flexibility, and planning [see Zohny (2015) for a review].

To adequately address the ethical, legal, and social implications of the use of neuroenhancement stimulant drugs at work, we briefly review specific evidence of beneficial and adverse effects of methylphenidate, amphetamine, and modafinil. We limit our discussion to these stimulants (Table 7.1) but recognize that other classes of drugs may become prevalent as neuroenhancers at work in the future, or they may be used in a subset of professions (e.g., erythropoietin in high-performance athletes, propranolol in elite musicians or the military) (Illes 2007).

7.2.1.1 Methylphenidate

Methylphenidate (commonly known under the brand name Ritalin) is a dopamine and noradrenaline reuptake inhibitor that acts by blocking the dopamine and noradrenaline transporters from gradually transporting used neurotransmitters back into the presynaptic neuron for reuse. In populations with attention deficit hyperactivity disorder (ADHD), methylphenidate allows neurotransmitters to act longer in the synapse by preventing reuptake, thereby increasing attention and concentration and decreasing reaction times and effects of fatigue [reviewed in Dubljevic (2013b)].

Table 7.1 Pharmacological neuroenhancers

Drug	Mechanism of action	Effect	Risk	Potential use in the workplace
Methylphenidate (e.g., Ritalin®, Concerta®)	Dopamine and noradrenaline reuptake inhibition (Dubljević 2013a)	Increases attention and concentration in populations with ADHD (Dubljević 2013b) Improves long-term memory when there is a long interval of time between the learning and recall in healthy adults (Repantis et al. 2010)	Tolerance (Dubljević and Ryan 2015) Psychological dependence, mania, and psychosis if abused (Dubljević 2013a)	Maintenance of baseline performance (Dubljević 2012b) Mitigation of fatigue (Dubljević 2013a) Enhanced speed of work (Dubljević 2013a)
Amphetamine (e.g., Adderall®, Dexedrine®, Vyvanse®)	Dopamine and norepinephrine reuptake inhibition (Zohny 2015) Monoamine oxidase (MAO) enzymes inhibition (Ranisch et al. 2013) Dopamine transporter reversal (Dubljević 2013a)	Promotes attention and concentration (Dubljević 2013a) Lacks evidence of reliable enhancement among healthy participants (Zohny 2015).	Addiction (Ranisch et al. 2013) Tolerance (Sharwood et al. 2013) Mania (STOA 2009) Psychosis (Dubljević 2013a) Disturbances of cognition, affect, and sensory experience (Zohny 2015)	Maintenance of baseline performance (Dubljević 2012b) Mitigation of fatigue (Ranisch et al. 2013) Enhanced speed of work (Dubljević 2013a)
Modafinil (e.g., Provigil®)	Unknown (Repantis et al. 2010)	Treats narcolepsy and sleep-deprivation-related fatigue (Dubljević 2016) Enhances executive functioning and memory for sleep-deprived individuals (Dubljević and Ryan 2015)	Safe in short-term use (Caldwell et al. 1999) Unknown risks for long-term use (Dubljević 2016)	Maintenance of baseline performance (Dubljević 2012a) Mitigation of fatigue (Estrada et al. 2012)

Repantis et al. (2010) found a significant improvement in long-term memory when there was a long interval of time between the learning phase and recall of study participants [see Faulmüller et al. (2014)], but they found no statistically significant improvements in attention, vigilance, or executive functions relevant to real-world performance (Dubljević and Ryan 2015).

Although moderate use of methylphenidate might be safe, chronic abuse can lead to tolerance and psychological dependence with varying degrees of abnormal behavior, such as mania and psychosis if injected or inhaled. It has been argued that time-release technology can effectively preclude non-oral use and danger of addiction, so extended-release formulas of methylphenidate might have a different danger profile than instant-release formulas (Dubljević 2013a).

7.2.1.2 Amphetamine

Like methylphenidate, amphetamines are used to treat individuals with ADHD with the goal of improving attention and concentration, but there is a key difference between the mechanisms of the two drugs. Amphetamine not only inhibits reuptake of dopamine and norepinephrine, but it also inhibits enzymes that inactivate and break down these neurotransmitters. It also has a specific effect that reverses the dopamine transporter action that increases further release of dopamine and norepinephrine into the synaptic cleft. Therefore, amphetamine is potentially more effective than methylphenidate but, at the same time, is associated with more risk factors, including addiction (Ranisch et al. 2013).

One study examining the effects of Adderall on 13 different measures of cognitive performance found no evidence of reliable enhancement among healthy participants as compared to ADHD patients. It did, however, find that the participants tended to believe that their cognitive performance was enhanced when taking Adderall (Zohny 2015). As for safety, there are valid concerns about the use of amphetamine. Namely, while legally prescribed amphetamine-based stimulant medications are less likely to cause physical dependence than illicit street drugs, they are nevertheless dangerous in terms of causing psychological dependence.

The potential for employers to coerce employees into enhancement drug use is not new (Dubljević 2016). For example, the self-reported prevalence of amphetamine use among truck drivers in Australia was reported to range between 19 and 32% (Sharwood et al. 2013). This has prompted policy makers to introduce security measures and to encourage whistle-blowing among employees who feel coerced into taking illegal drugs. With the imposition of random roadside drug testing (RTA 2009), the prevalence of use dropped to 3.9% (Sharwood et al. 2013). However, such regulatory measures are costly and limited to certain drugs to which they can be applied, such as to amphetamines, as well as cannabis, alcohol, cocaine, and opiates (Medical Bureau of Roadside Safety 2012).

7.2.1.3 Modafinil

Modafinil, a drug developed for the treatment of narcolepsy and sleep-deprivation-related fatigue, has low toxicity and is believed to pose only modest short-term risks when used for nonmedical, enhancement-related purposes (Caldwell et al. 1999). Indeed, empirical studies conducted on healthy adults for the military suggest the beneficial replacement of amphetamine with modafinil and for use in combat missions (Estrada et al. 2012).

Authors of a recent review of single-dose studies of modafinil concluded that modafinil enhances executive functioning and memory for sleep-deprived individuals but less reliably for fully rested adults (Dubljević and Ryan 2015). There were also positive findings for specific tasks, such as those that required inhibitory control, and occasional impairment effects on other tasks.

Given safety profile and performance maintenance effects, modafinil presents a promising option for workplace enhancement. However, Dubljević (2012a, b) described a phenomenon concerning "truckers on modafinil" that illustrates the profound dangers of allowing corporate actors to pursue positional advantage without regulation (Dubljević 2012a). If modafinil is not regulated appropriately, it might produce an overall increase in different forms of shift work and related health (e.g., cardiovascular disease, increased mortality), social (e.g., work-life balance, poor parenting), and organizational (e.g., burnout, turnover, absenteeism) costs (Dubljević 2016). Further, the expansion of the modafinil label to include treatment of shift work sleep disorder might lead to a change in social practices and produce drastic effects in the workforce by normalizing an otherwise exceptional condition of working during the night.

7.2.2 Neurotechnological Enhancement

Beyond pharmaceutical enhancement, people are seeking other forms of neuro-enhancement to boost performance and brain function. The neurotechnology market is expected to grow to $12.22 billion by 2021 (Markets and Markets 2017), and part of that market includes transcranial direct-current stimulation (tDCS), transcranial magnetic stimulation (TMS), and electroencephalographic (EEG)-based devices that are increasingly sold and unregulated on the direct-to-consumer commercial market. The claims made about these devices are varied (Coates McCall and Illes 2018), and many companies advertise beneficial effects on cognitive functioning, including alertness, concentration, mathematical ability, and even physical athleticism. These devices present the possibility of neuroenhancement without the need for a prescription or illegal purchase, hence removing at least one barrier to access (Table 7.2).

7.2.2.1 Transcranial Direct-Current Stimulation

Direct-to-consumer transcranial direct-current stimulation (tDCS) devices are an exploding business. They are marketed to increase memory, boost creativity, and augment physical strength and athleticism (Dubljević et al. 2014; Wurzman et al. 2017; Fitz and Reiner 2013). It is not hard to imagine employers adopting such devices for employee use. tDCS works to change brain functioning by causing the resting membrane potential of neurons in the brain to depolarize or hyperpolarize. The membrane potential of a neuron determines its propensity to fire an action potential, send electricity along its length, and release chemical signalers to other cells—the basic mode of brain communication and functioning. When a tDCS device delivers positive stimulation from its anodal electrode, the current depolarizes

Table 7.2 Device-based neuroenhancers

Device	Mechanism of action	Effect	Risk	Potential use in the workplace
tDCS	Depolarization of neuron resting membrane potential by electrical stimulation (Nitsche and Paulus 2000)	Improves working memory (Fregni et al. 2005; Kuo and Nitsche 2012; Teo et al. 2011; Zaehle et al. 2011) Trains fine motor skills (Williams et al. 2010) Enhances executive function (Miler et al. 2017) Increases performance endurance (Angius et al. 2018) Heightens vigilance (Clark et al. 2012; Horvath et al. 2015a, b)	Skin irritation (Wurzman et al. 2017) Nausea (Fitz and Reiner 2013) Headache (Dubljević 2015a) Phosphene formation (Utz et al. 2010)	Heightened vigilance (Clark et al. 2012) Enhanced cognitive functioning (Fregni et al. 2005) Increased productivity (Kuo and Nitshce 2012) Memory formation (Utz et al. 2010) Enhanced creativity (Wexler 2016) Improved job training (Williams et al. 2010)
TMS	Increased excitability of neurons caused by current induction using high-intensity magnetic fields (Grosbras and Paus 2003, 2002; Töpper et al. 1998)	Improves perceptual, motor, and executive processing (Luber and Lisanby 2014) Improves perceptual discrimination (Grosbras and Paus 2003) Improves motor learning (STOA 2009) Enhances recognition memory (Turriziani 2012) Improves planning and executive function (Marron et al. 2017) Increases speed of psychomotor performance (Bliss et al. 2003) Increases speed of visual search and object identification (Martin et al. 2017) Causes superior performance on	Syncope (Dubljević 2015a) Induced seizures (STOA 2009) Discomfort or pain (Snyder 2009) Transient induction of hypomania (Dubljević and Racine 2017) Transient cognitive changes (Snyder et al. 2006) Transient hearing loss (Gersner et al. 2016) Transient impairment of working	Improved job training through long-term potentiation (Bliss et al. 2003)

(continued)

Table 7.2 (continued)

Device	Mechanism of action	Effect	Risk	Potential use in the workplace
		attention, memory, and language tasks (Luber and Lisanby 2014)	memory (Berlim et al. 2013)	
EEG	Electrophysiological monitoring measures voltage fluctuations resulting from ionic currents within neurons	Measures level of cognitive arousal (Teplan 2002) Trains the brain to improve concentration (Dubljević 2015a)	No known physical risks Potential for surveillance (Tekabi et al. 2016)	Detection of fatigue (Teplan 2002) Increased safety (SmartCap Technologies) Neural control of machines (Millán et al. 2004)

the resting membrane potential and increases neuronal excitability and spontaneous firing of neurons (Nitsche and Paulus 2000). The effect of this stimulation on cortical neurons can persist even after the stimulation ends. How long this change lasts depends on the length and strength of the stimulation delivered; the longer the stimulation and the higher the strength of stimulation, the longer the effects last after stimulation ends (Utz et al. 2010). Following safety protocols, tDCS is widely regarded as safe for human use; however these devices remain unregulated and have not been approved for any clinical indication by the FDA in the United States (Wexler 2016).

Preliminary evidence has suggested that tDCS may have functional effects in healthy humans and that it has the potential to induce significant alteration of functional connectivity (Polanía et al. 2011; Zaehle et al. 2011). In healthy adults, CE-certified tDCS devices have been demonstrated to be successful in promoting working memory (Fregni et al. 2005; Kuo and Nitshce 2012; Teo et al. 2011; Zaehle et al. 2011), fine motor skills training (Williams et al. 2010), executive attention (Miler et al. 2017), endurance performance (Angius et al. 2018), and heightened vigilance (Clark et al. 2012; Horvath et al. 2015a, b). By contrast, a 2017 analysis found no evidential value of enhancement in samples of tDCS studies of cognition and working memory in healthy populations (Medina and Cason 2017). One study actually found that a commercially available tDCS stimulation device significantly decreased user accuracy on a working memory task (Steenbergen et al. 2016).

It is not difficult to foresee ways in which such devices may seem desirable to employers and how they might impact the workplace. They are relatively small, and many of them are completely wireless, allowing for use in a variety of situations and environments. Their use could heighten vigilance for safer workplaces, including the potential for alertness to safety threats before an accident could happen. Enhanced cognitive functioning could increase productivity in knowledge-based industries. Promoting memory formation could decrease training times for new employees or when retraining older employees to use new equipment. Enhanced creativity could lead to innovation.

As with any cognitive enhancement, however, wearable tDCS devices present the potential for coerced use. In the drive for constant improvements in productivity, brain stimulation that makes a person work faster and be smarter for prolonged periods of time could become mandated, or at least expected. This raises ethical concerns about employee autonomy and fairness. The increased productivity and creativity would stand to primarily benefit the employer, while the health risks associated with the use of the device would be borne solely by the employees.

7.2.2.2 Transcranial Magnetic Stimulation

Transcranial magnetic stimulation (TMS) uses very brief bursts of high-intensity magnetic fields to induce electrical currents in the neurons in a small region of the cortex under a magnetic coil. Similar to tDCS, the neural effects of TMS depend on the frequency and duration of stimulation. If the frequency of TMS stimulation is 1 Hz or greater, the stimulation is known as repetitive TMS (rTMS). On the low-frequency end of rTMS (about 1 Hz), cortical excitability decreases. Contrary to that, higher-frequency rTMS (5–20 Hz) increases cortical excitability. The timing of TMS relative to a given task can also influence whether the stimulation is disruptive or facilitatory (Snyder 2009). This ability to upregulate or downregulate cortical excitability—and the ability to vary the degree of stimulation provided—makes TMS a good potential candidate for altering cognitive performance through the manipulation of cortical networks.

TMS appears to achieve its enhancing effects through one of the following two mechanisms: first, by directly modulating a cortical region or network that leads to more efficient processing and, second, by disrupting the normal processes that compete with or distract from task performance, termed "addition-by-subtraction" (Luber and Lisanby 2014). Clinically, TMS is used to diagnose conditions such as amyotrophic lateral sclerosis, multiple sclerosis, and stroke (Groppa et al. 2012). As a therapy, it is used for a wide range of conditions such as Alzheimer's disease (Rutherford et al. 2015), epilepsy (Gersner et al. 2016), and treatment-resistant depression (Berlim et al. 2013).

More than 60 studies have reported significant improvements of behavioral and cognitive functions after TMS, including the speed and accuracy of perceptual, motor, and executive processing tasks (Luber and Lisanby 2014). These also include reports of better perceptual discrimination (Grosbras and Paus 2003), improved motor learning, enhanced recognition memory (Turriziani 2012), improved planning abilities (Marron et al. 2017), speedier visual search and object identification (Martin et al. 2017), and superior performance on attention, memory, and language tasks (Luber and Lisanby 2014). One study reported inducing savant-like ability with numbers in otherwise ordinary healthy people (Snyder et al. 2006). Most of these enhancement effects, usually reflected in decreased reaction times, result from direct TMS to specific task-related brain regions (Dubljević and Racine 2017) and suggest that the stimulation potentiates local neural activity (Grosbras and Paus 2003, 2002; Töpper et al. 1998). The persistence of effects beyond the end of the stimulation also suggests that TMS may additionally exert its effects through the strengthening of synapses known as long-term potentiation and through the changing of the synaptic plasticity of relevant brain regions (Bliss et al. 2003).

Effects of TMS can be online (occurring during the stimulation itself) or offline (occurring after the stimulation has ended). Since TMS stimulation at the time of the task to be enhanced seems unlikely in a workplace environment since the device itself is relatively large, hardwired, and cumbersome, TMS enhancement offline, such as for learning and skill acquisition, alone could be instrumental in job training. Further, given that TMS devices are not portable, it is unlikely that TMS will be on the market as a product per se, but rather, it may be used as a service (Dubljević 2015a).

7.2.2.3 Electroencephalography (EEG)-Based Devices

Although neuroenhancement usually carries with it a connotation of direct physical intervention in the brain, more indirect technologies can also lead to improved cognitive functioning. While not directly stimulating in the way that cognitive-enhancing drugs or tDCS are, EEG can be seen as neuroenhancement in the workplace by way of its use for increased productivity. Portable wireless EEG headsets are being sold directly to consumers today for the purpose of training the brain and improving cognitive functioning. EEG can inform a user of her or his level of arousal, alertness, and stress through universally recognizable patterns of cortical activity (Teplan 2002). By recognizing and alerting the user to mental arousal at a given time, neurofeedback mobilizes mental resources and focus on command. Many companies are developing these devices specifically for the workplace to help increase safety in jobs that require sustained alertness. For example, the Australian company SmartCap Technologies has developed a trucker's cap that it claims can detect drowsiness. The signal triggers a smartphone alarm or alerts a supervisor or dispatcher if the driver is in danger (SmartCap Technologies). Automaker Nissan has also developed a brain-to-vehicle (B2V) computer interface to reduce driver reaction time by up to half a second (Nissan Motor Co.). Similar devices have been proposed for workers who operate heavy machinery: the EEG signal alerts them if they grow drowsy so that they can be taken off the floor. EEG could even be used in office settings to track if employees become distracted over the course of the day. It is not inconceivable that, in the future, employers may devise a scheme to pay their employees only for the number of alert hours worked.

Proponents claim that continuously and unobtrusively monitoring levels of task engagement and mental workload in a work environment could help identify more efficient methods for employees to interact with technology (Berka et al. 2007). Again, while EEG devices are not providing stimulation, these products can be seen as neuroenhancement because they are neurotechnologies that augment human functioning and increase productivity. Alternatively, the constant monitoring of levels of engagement during work hours could instead be used to punish those who get distracted, are tired, or are not focused or alert. Neural surveillance may thus contravene freedom, liberty, and privacy; sow seeds of distrust or contempt (Ball 2010); and create a toxic work environment.

7.3 Regulatory Options for Neuroenhancement

In this section, we focus on neuroenhancement in the context of its ethical and societal importance. The neuroenhancement phenomenon is a result of a mixture of individual choices, organizational needs, and social preferences. Consequently, three different levels of decision-making (private, nonpublic, and public) have to be taken into consideration when assessing the quality of arguments marshaled in the debate, at least in functioning democracies.

Private reasons constrained by rights of others are the basis for the most permissive perspectives. Indeed, a strong concept of freedom grounds the call for respecting even irrational choices of individuals. That is, even in the absence of any cognitive-enhancing effect, a personal preference for the recreational use of substances is a driving force for social uptake.

Group or nonpublic decision-making is often constrained by the shared values of a group [see Rawls (1997)]. Nonpublic decision-making represents the interests of a group or subset of society, and it is not problematic as long as the specific values are not coercively enforced on those outside the group and the interests of the public are not encroached upon. For example, a corporation has the right to impose drug tests on its employees as a requirement for job retention as long as it is conducted in full compliance with the law. A group can be ideological (e.g., a religious community) or based on mutual interest (e.g., a corporation) or both. The key point is that individuals—at least in democratic societies—have a choice of either conforming to the values and norms of a group, reaping the benefits of membership or leaving the group, thus losing associated gains. The norms of the group could be opposed to prevailing social norms (e.g., maximizing profits at the expense of the environment) or aligned with them, extending shared social values into demanding ethical requirements (e.g., fair trade movement).

Public decision-making applies to values that are either shared by all of society or norms that are necessary for the continued functioning of a fair system of social cooperation. Public decision-making requires democracy and vice versa. Democracy assumes the need for public justification and fallibility in every decision and judgment, no matter what the expertise of the decision-makers may be. Public justification can only be based on proper use of reason and democratic dialogue. Democratic dialogue requires that any given justification can be criticized, and any rules can, in principle, be changed if they promote self-serving interest of a particular group as opposed to public interest. Thus, rules and norms that apply to all must be justified in a transparent manner.

Regulatory options that are available in democratic societies mirror ethical assessments (Dubljević 2012b). If a practice is widely seen to be morally unacceptable (e.g., selling cocaine in the workplace) it results in prohibitionist policies, morally bad but still acceptable behavior (e.g., drinking excessive amounts of alcohol) results in discourage-use policies, morally irrelevant behaviors (e.g., drinking coffee) result in laissez-faire approaches, morally good behavior that is not obligatory (e.g., studying for a university degree) results in encourage-use policies, and morally obligatory practices (e.g., education) result in mandatory policies.

The case of regulation of neuroenhancement, then, depends on the level of nonpublic and public decision-making and social consensus on the moral status of the practice. Currently, there is no consensus on the moral status of neuroenhancement. The academic debate is not resolved, and available empirical evidence on the social attitudes about the practice is ambiguous. In Australia, 85% of a sample of the general population believe that the use of medications for neuroenhancement is morally unacceptable (Partridge et al. 2012). Similar findings hold for populations surveyed in the United States (Dodge et al. 2012) and Germany (Dubljević et al. 2014). Other surveys in the United States report that neuroenhancement is viewed as acceptable and that it does not represent an unfair advantage [for a review, see Dubljević (2015b)].

This variability gives employers and companies much leeway. According to critics of neuroenhancement, through competitive pressure and the desire to gain at the expense of others, contagion processes would start to affect different dimensions of the basic structure of society in the long run (Dubljević 2013b; Racine and Forlini 2009). Thus, a discourage-use policy is most likely to be publicly justifiable. Discouragement could take multiple forms, including taxation, as described by the *Economic Disincentives Model (EDM)* (Dubljević 2012b). The EDM requires government regulation of the cost of neuroenhancers through the imposition of taxes, fees, medical tests for users, and additional insurance requirements. Standard costs of production and distribution would be incorporated into the price of commercially available enhancers, while the profit margin would be limited and additional taxes imposed. Companies that earn profits from neuroenhancers or that encourage the use of cognitive enhancers among their employees would be further taxed. Similar to the allocation of tobacco-related taxes, funds collected through the EDM could be used toward healthcare for members of the population for whom accessibility is limited (e.g., those living below poverty line) and the remaining funds allocated to allow education access to marginalized groups, such as minorities and aboriginal peoples.

The strengths of this model are (1) state neutrality on personal preferences, (2) protection of the best interests of all citizens, (3) reliable data on consumption and demand, and (4) effective evaluation of long-term health costs among users of neuroenhancement. However, the weaknesses of the EDM lie in its potential to increase financial burdens on employers and employees, inequality in the workplace, and harm to those the policy seeks to protect. For example, if truck drivers are coerced into taking modafinil by their employers, any downstream financial incentives that may accrue to the company are offset by upstream costs. Moreover, coercion equates to exploitation, and exploitation threatens employee autonomy, especially the autonomy of people who may already be disenfranchised. A significant regulatory gap remains to be filled.

7.4 Conclusion

Stimulant drugs such as methylphenidate, amphetamine, and modafinil have already found their way into the work environment. Devices that modulate and record brain function are reinvigorating older discourse around the ethics of pharmacologic enhancement and can be expected to play an increasing role as the direct-to-consumer neurotechnology market grows. With products claiming to boost memory and creativity among other valuable human performance functions, it is easy to imagine that employers may seek to adopt such devices for use by their employees. Indeed, the push toward the incorporation of wearable technologies in the workplace can be seen in employee health programs that seek to lower company healthcare costs and boost employee efficiency and productivity (Pettey 2018). We predict that neurotechnological enhancement will likely be received with the same enthusiasm.

More research is needed to provide evidence about the safety, efficacy, and social implications of all forms of enhancement within and outside of the workplace. Public policies must be informed both by science and by the views of stakeholders who possess a wide range of skill sets and represent diverse cultural and socioeconomic backgrounds. Indeed, in a regulatory environment in which benefits, harms, and justice are embraced and carefully balanced, neuroenhancers may be positively integrated into workplace settings in the future.

Acknowledgements VD is supported by the Faculty Research and Professional Development program at NC State University. This project was also supported in part by Technical Safety BC, Vancouver, Canada (ICM and JI) and Neuroethics Canada. JI is Canada Research Chair in Neuroethics. Special thanks go to Abigail Scheper for her help with language editing and formatting.

References

Academy of Medical Sciences. Human enhancement and the future of work. In: Joint report of the Academy of Medical Sciences, the British Academy, the Royal Academy of Engineering and the Royal Society. 2012. http://royalsociety.org/upload-edFiles/Royal_Society_Content/policy/projects/human-enhancement/2012-11-06-Human-enhancement.pdf. Accessed 10 Nov 2012.

Angius L, Mauger AR, Hopker J, Pascual-Leone A, Santarnecchi E, Marcora SM. Bilateral extracephalic transcranial direct current stimulation improves endurance performance in healthy individuals. Brain Stimul. 2018;11(1):108–17.

Appel JM. When the boss turns pusher: a proposal for employee protections in the age of cosmetic neurology. J Med Ethics. 2008;34(8):616–8.

Ball K. Workplace surveillance: an overview. Labor Hist. 2010;51(1):87–106.

Berka C, Levendowski DJ, Lumicao MN, Yau A, Davis G, Zivkovic VT, Olmstead RE, Tremoulet PD, Craven PL. EEG correlates of task engagement and mental workload in vigilance, learning, and memory tasks. Aviat Space Environ Med. 2007;78(5):B231–44.

Berlim MT, Van den Eynde F, Daskalakis ZJ. Clinically meaningful efficacy and acceptability of low-frequency repetitive transcranial magnetic stimulation (rTMS) for treating primary major depression: a meta-analysis of randomized, double-blind and sham-controlled trials. Neuropsychopharmacology. 2013;38(4):543.

Bliss TV, Collingridge GL, Morris RG. Introduction. Long-term potentiation and structure of the issue. Philos Trans R Soc B. 2003;358(1432):607.

British Medical Association. Boosting your brainpower: ethical aspects of cognitive enhancements. A discussion paper from the British Medical Association. London: BMA; 2007.

Caidwell Jr JA, Smythe III NK, Caidwell J, Hall KK, Norman DN. The effects of modafinil on aviator performance during 40 hours of continuous wakefulness: a UH-60 helicopter simulator study. Army Aeromedical Research Unit Fort Rucker Al; 1999. Report No.: 99-17.

Clark VP, Coffman BA, Mayer AR, Weisend MP, Lane TD, Calhoun VD, Raybourn EM, Garcia CM, Wassermann EM. TDCS guided using fMRI significantly accelerates learning to identify concealed objects. NeuroImage. 2012;59(1):117–28.

Coates McCall I, Illes J. Owning ethical innovation: claims about commercial brain wearable technologies. International Neuroethics Society Annual Meeting. 2018. Abstract submitted for publication.

Dodge T, Williams KJ, Marzell M, Turrisi R. Judging cheaters: is substance misuse viewed similarly in the athletic and academic domains? Psychol Addict Behav. 2012;26(3):678–82.

Dubljević V. Principles of justice as the basis for public policy on psychopharmacological cognitive enhancement. Law Innov Technol. 2012a;4(1):67–83.

Dubljević V. Toward a legitimate public policy on cognition-enhancement drugs. AJOB Neurosci. 2012b;3(3):29–33.

Dubljević V. Prohibition or coffee shops: regulation of amphetamine and methylphenidate for enhancement use by healthy adults. Am J Bioeth. 2013a;13(7):23–33. https://doi.org/10.1080/15265161.2013.794875.

Dubljević V. Cognitive enhancement, rational choice and justification. Neuroethics. 2013b;6(1):179–87.

Dubljević V. Neurostimulation devices for cognitive enhancement: toward a comprehensive regulatory framework. Neuroethics. 2015a;8(2):115–26.

Dubljević V. Cognitive enhancement: a glance at the future and ethical considerations. In: Knafo S, Venero C, editors. Cognitive enhancement. Amsterdam: Elsevier; 2015b. p. 343–65.

Dubljević V. Enhancement with modafinil: benefiting or harming the society? In: Jotterand F, Dubljević V, editors. Cognitive enhancement: ethical and policy implications in international perspectives. New York: Oxford University Press; 2016. p. 259–74.

Dubljević V, Racine E. Moral enhancement meets normative and empirical reality: assessing the practical feasibility of moral enhancement neurotechnologies. Bioethics. 2017;31(5):338–48.

Dubljević V, Ryan CJ. Cognitive enhancement with methylphenidate and modafinil: conceptual advances and societal implications. Neurosci Neuroecon. 2015;4:25–33. https://doi.org/10.2147/NAN.S61925.

Dubljević V, Sattler S, Racine E. Cognitive enhancement and academic misconduct: a study exploring their frequency and relationship. Ethics Behav. 2014;24(5):408–20.

Enriquez-Geppert S, Huster RJ, Herrmann CS. EEG-neurofeedback as a tool to modulate cognition and behavior: a review tutorial. Front Hum Neurosci. 2017;11:51.

Estrada A, Kelley AM, Webb CM, Athy JR, Crowley JS. Modafinil as a replacement for dextroamphetamine for sustaining alertness in military helicopter pilots. Aviat Space Environ Med. 2012;83(6):556–64.

Faulmüller N, Malsen H, Savulescu J. Pharmacological cognitive enhancement—how neuroscientific research could advance ethical debate. Front Syst Neurosci. 2014;8:1–12. https://doi.org/10.3389/fnsys.2014.00107.

Fitz NS, Reiner PB. The challenge of crafting policy for do-it-yourself brain stimulation. J Med Ethics. 2013;41:410–2.

Fregni F, Boggio PS, Nitsche M, Bermpohl F, Antal A, Feredoes E, Marcolin MA, Rigonatti SP, Silva MT, Paulus W, Pascual-Leone A. Anodal transcranial direct current stimulation of prefrontal cortex enhances working memory. Exp Brain Res. 2005;166(1):23–30.

Gersner R, Oberman L, Sanchez MJ, Chiriboga N, Kaye HL, Pascual-Leone A, Libenson M, Roth Y, Zangen A, Rotenberg A. H-coil repetitive transcranial magnetic stimulation for treatment of temporal lobe epilepsy: a case report. Epilepsy Behav Case Rep. 2016;5:52–6.

Groppa S, Oliviero A, Eisen A, Quartarone A, Cohen LG, Mall V, Kaelin-Lang A, Mima T, Rossi SE, Thickbroom GW, Rossini PM. A practical guide to diagnostic transcranial magnetic stimulation: report of an IFCN committee. Clin Neurophysiol. 2012;123(5):858–82.

Grosbras MH, Paus T. Transcranial magnetic stimulation of the human frontal eye field: effects on visual perception and attention. J Cogn Neurosci. 2002;14(7):1109–20.

Grosbras MH, Paus T. Transcranial magnetic stimulation of the human frontal eye field facilitates visual awareness. Eur J Neurosci. 2003;18(11):3121–6.

Horvath JC, Forte JD, Carter O. Quantitative review finds no evidence of cognitive effects in healthy populations from single-session transcranial direct current stimulation (tDCS). Brain Stimul. 2015a;8(3):535–50.

Horvath JC, Forte JD, Carter O. Evidence that transcranial direct current stimulation (tDCS) generates little-to-no reliable neurophysiologic effect beyond MEP amplitude modulation in healthy human subjects: a systematic review. Neuropsychologia. 2015b;66:213–36.

Illes J. Not forgetting. Am J Bioeth. 2007;7(9):3–4.

Illieva I, Boland J, Farah MJ. Objective and subjective cognitive enhancing effects of mixed amphetamine salts in healthy people. Neuropharmacology. 2013;64:496–505.

Kuo MF, Nitsche MA. Effects of transcranial electrical stimulation on cognition. Clin EEG Neurosci. 2012;43(3):192–9.

Luber B, Lisanby SH. Enhancement of human cognitive performance using transcranial magnetic stimulation (TMS). NeuroImage. 2014;85:961–70.

Markets and Markets. Brain monitoring market by product (MRI, CT, PET, EEG, EMG, MEG, TCD, ICP, Electrodes, Sensors, Gels, Cables), procedure (invasive, non-invasive), disease (TBI, stroke, dementia, epilepsy) & end user (hospital, clinic, ASC, ambulance)—forecasts to 2021. 2017. https://www.marketsandmarkets.com/Market-Reports/brain-monitoring-devices-market-909.html?gclid=EAIaIQobChMIiPqg1e7l2wIVBGt-Ch1jYQewEAAYASAAEgLf6_D_BwE. Accessed 21 Jun 2018.

Marron EM, Viejo-Sobera R, Palaus M, Boixadós M, Valero-Cabre A, Redolar-Ripoll D. P237 Modulating executive functions and working memory performance on clinical neuropsychological tasks with theta burst transcranial magnetic stimulation. Clin Neurophysiol. 2017;128 (3):e130–1.

Martin DM, McClintock SM, Forster JJ, Lo TY, Loo CK. Cognitive enhancing effects of rTMS administered to the prefrontal cortex in patients with depression: a systematic review and meta-analysis of individual task effects. Depress Anxiety. 2017;34(11):1029–39.

Medical Bureau of Roadside Safety. Report on roadside drug testing and equipment and related matters. Dublin: Medical Bureau of Roadside Safety; 2012.

Medina J, Cason S. No evidential value in samples of transcranial direct current stimulation (tDCS) studies of cognition and working memory in healthy populations. Cortex. 2017;94:131–41.

Miler JA, Meron D, Baldwin DS, Garner M. The effect of prefrontal transcranial direct current stimulation on attention network function in healthy volunteers. Neuromodulation. 2017;21 (4):355–61.

Millán JR, Renkens F, Mourino J, Gerstner W. Noninvasive brain-actuated control of a mobile robot by human EEG. IEEE Trans Biomed Eng. 2004 Jun;51(6):1026–33.

NISSAN MOTOR Co., Ltd. Brain-to-Vehicle [Internet]. NISSAN|CORPORATE INFORMATION|Outline of Company TOP. [cited 2018 May 16]. Available from https://www.nissan-global.com/EN/TECHNOLOGY/OVERVIEW/b2v.html.

Nitsche MA, Paulus W. Excitability changes induced in the human motor cortex by weak transcranial direct current stimulation. J Physiol. 2000;527(3):633–9.

Partridge B, Lucke J, Hall W. A comparison of attitudes toward cognitive enhancement and legalized doping in sport in a community sample of Australian adults. Am J Bioeth Primary Res. 2012;3(4):81–6.

Pettey C. Wearables hold the key to connected health monitoring [Internet]. Gartner.com. 2018 [cited 2018 Oct 11]. Available from https://www.gartner.com/smarterwithgartner/wearables-hold-the-key-to-connected-health-monitoring/

Polanía R, Nitsche MA, Paulus W. Modulating functional connectivity patterns and topological functional organization of the human brain with transcranial direct current stimulation. Hum Brain Mapp. 2011;32(8):1236–49.

Racine E, Forlini C. Expectations regarding cognitive enhancement create substantial challenges. J Med Ethics. 2009;35(8):469–70.

Ranisch R, Garofoli D, Dubljević V. 'Clock shock', motivational enhancement and performance maintenance in adderall use. AJOB Neurosci. 2013;4(1):13–4.

Rawls J. The idea of public reason revisited. Univ Chic Law Rev. 1997;64(3):765–807.

Repantis D, Schlattmann P, Laisney O, Heuser I. Modafinil and methylphenidate for neuroenhancement in healthy individuals: a systematic review. Pharmacol Res. 2010;62:187–206.

Roads and Track Authority [RTA]. Roadside drug testing. Sydney: New South Wales Centre for Road Safety. 2009. http://www.rta.nsw.gov.au/heavyvehicles/downloads/hv_drug_testing_dl1.html. Accessed 2 May 2013.

Rutherford G, Lithgow B, Moussavi Z. Short and long-term effects of rTMS treatment on Alzheimer's disease at different stages: a pilot study. J Exp Neurosci. 2015;9:JEN-S24004.

Science and Technology Options Assessment [STOA]. Human enhancement study. The Hague: Rathenau Institute; 2009.

Sharwood LN, Elkington J, Meuleners L, Ivers R, Boufous S, Stevenson M. Use of caffeinated substances and risk of crashes in long distance drivers of commercial vehicles: case-control study. BMJ. 2013;346:f1140. https://doi.org/10.1136/bmj.f1140.

SmartCap Technologies. SmartCap Technologies|Measure Alertness. Eliminate Fatigue [Internet]. SmartCapTech. [cited 2018 Jun 11]. Available from https://www.smartcaptech.com/

Snyder A. Explaining and inducing savant skills: privileged access to lower level, less-processed information. Philos Trans R Soc B 2009;364(1522):1399–405. Accessed 11 Apr 2013.

Snyder A, Bahramali H, Hawker T, Mitchell DJ. Savant-like numerosity skills revealed in normal people by magnetic pulses. Perception. 2006;35(6):837–45.

Steenbergen L, Sellaro R, Hommel B, Lindenberger U, Kühn S, Colzato LS. "Unfocus" on foc. us: commercial tDCS headset impairs working memory. Exp Brain Res. 2016;234(3):637–43.

Takabi H, Bhalotiya A, Alohaly M. Brain computer interface (BCI) applications: privacy threats and countermeasures. In Collaboration and Internet Computing (CIC), 2016 IEEE 2nd International Conference. 2016 Nov 1. IEEE. p. 102–11.

Teo F, Hoy KE, Daskalakis ZJ, Fitzgerald PB. Investigating the role of current strength in tDCS modulation of working memory performance in healthy controls. Front Psychiatry. 2011;2:45.

Teplan M. Fundamental of EEG measurement. Meas Sci Rev. 2002;2(2):1–11.

Töpper R, Mottaghy FM, Brügmann M, Noth J, Huber W. Facilitation of picture naming by focal transcranial magnetic stimulation of Wernicke's area. Exp Brain Res. 1998;121(4):371–8.

Turriziani P. Enhancing memory performance with rTMS in healthy subjects and individuals with mild cognitive impairment: the role of the right dorsolateral prefrontal cortex. Front Hum Neurosci. 2012;6:62.

Utz KS, Dimova V, Oppenländer K, Kerkhoff G. Electrified minds: transcranial direct current stimulation (tDCS) and galvanic vestibular stimulation (GVS) as methods of non-invasive brain stimulation in neuropsychology—a review of current data and future implications. Neuropsychologia. 2010;48(10):2789–810.

Wexler A. A pragmatic analysis of the regulation of consumer transcranial direct current stimulation (TDCS) devices in the United States. J Law Biosci. 2016;2(3):669–96.

Williams JA, Pascual-Leone A, Fregni F. Interhemispheric modulation induced by cortical stimulation and motor training. Phys Ther. 2010;90(3):398–410.

Wurzman R, Hamilton R, Pascual-Leone A, Fox M. An open letter concerning do-it-yourself (DIY) users of transcranial direct current stimulation (tDCS). Ann Neurol. 2017;80(1):1–4. https://doi.org/10.1002/ana.24689.

Zaehle T, Sandmann P, Thorne JD, Jäncke L, Herrmann CS. Transcranial direct current stimulation of the prefrontal cortex modulates working memory performance: combined behavioural and electrophysiological evidence. BMC Neurosci. 2011;12(1):2.

Zohny H. The myth of cognitive enhancement drugs. Neuroethics. 2015;8:257–69. https://doi.org/10.1007/s12152-015-9232-9.

Part II

The Neuroscience of Organizational Ethics

Section Introduction: The Neuroscience of Organizational Ethics

8

Joé T. Martineau and Eric Racine

Contents

The idea that neuroscience has the potential to enrich and have fundamental implications for ethics is not new but rather part of an ongoing debate about the contribution of neuroscience to the humanities. Already in 1967, physician and neuroscientist Paul MacLean wrote that understandings of the neuronal bases of empathy had the potential to change our view of moral decision-making and empathy in medicine (MacLean 1967). However, alongside celebrations of the promises of neuroscience and its epistemic supremacy in the domain of ethics (Gazzaniga 2005; Churchland 2008), passionate critiques of this view have emerged, deemed as "reductionist" by social scientists and ethicists [for a review, see Pickersgill (2013) and Racine (2013)]. Between an anti-naturalist rejection of the potential of neuroscience to contribute to ethics and the opposite strong positivist sweeping assertions that neuroscience will revolutionize the way we consider morality and offer new foundations to ethics as a discipline, a series of middle-ground moderate naturalistic positions have been proposed (Racine et al. 2017), including those inspired by Dewey's pragmatic philosophy and his thinking on the

J. T. Martineau (✉)
Department of Management, HEC Montréal, Montreal, QC, Canada
e-mail: joe.trempe-martineau@hec.ca

E. Racine
IRCM, Montréal, QC, Canada

Université de Montréal, Montréal, QC, Canada

McGill University, Montréal, QC, Canada
e-mail: eric.racine@ircm.qc.ca

J. T. Martineau, E. Racine (eds.), *Organizational Neuroethics*, Advances in Neuroethics, https://doi.org/10.1007/978-3-030-27177-0_8

role of evidence in ethics (Racine 2010). In our eyes, claims about the foundational nature of neuroscience knowledge, i.e., that a "brain-based ethics" could be derived from the neuroscience of morality, are misled and involve different types of logical fallacies (Racine 2013). However, neuroscience undeniably has important contributions to make to the field of ethics but only if these are brought into dialogue with perspectives from other disciplines as a way to support deliberative processes about different ethical issues.

This second part of the book features seven chapters. Robertson explores how research in decision neuroscience contributes to organizational ethics and how business ethicists can themselves contribute to the development of neuroscience research. Craze provides a critical analysis of corporate social responsibility initiatives, based on neuroscience evidence, and on how in its current conceptualizations corporate social responsibility can lead to dehumanization. Lozano tackles the topic of ethical and unethical decision-making in organizations, integrating insights from neuroscience and cognitive science. We the editors (Martineau and Racine) alongside Decety review research on the social neuroscience of empathy and its implications for business ethics. Karmarkar and Jenkins address how neuroscience research on trust and uncertainty contributes to our understanding of online interactions. Ewusi-Boisvert offers an original perspective on anger expression in the workplace, highlighting the potential positive contributions of moral anger, based on neuroscience evidence. Finally, Tachibana explores how space neuroscience research can inform performance management and business ethics.

References

Churchland PS. The significance of neuroscience for philosophy. Funct Neurol. 2008;23(4):175–8.

Gazzaniga MS. The ethical brain. New York: Dana Press; 2005.

MacLean PD. The brain in relation to empathy and medical education. J Nerv Ment Dis. 1967;144 (5):374–82.

Pickersgill M. The social life of the brain: neuroscience in society. Curr Sociol. 2013;61(3):322–40.

Racine E. Pragmatic neuroethics. Improving treatment and understanding of the mind brain. Cambridge, MA: MIT Press; 2010.

Racine E. Pragmatism and the contribution of neuroscience to ethics. Essays Philos Hum. 2013;21 (1):13–30.

Racine E, Dubljevic V, Jox RJ, Baertschi B, Christensen JF, Farisco M, et al. Can neuroscience contribute to practical ethics? A critical review and discussion of the methodological and translational challenges of the neuroscience of ethics. Bioethics. 2017;31(5):328–37.

Decision Neuroscience and Organizational Ethics

9

Diana C. Robertson

Contents

Abstract

Recent advances in decision neuroscience hold promise to address questions that have engaged organizational ethics scholars for decades. Following from this promise, two research questions emerge: (1) What are the most productive ways to enhance organizational ethics through scholarly communication and collaborative work among philosophers, social scientists, and neuroscientists? (2) What contributions have already been made, and how can we build on those contributions? Specifically, how can scholars in the field of organizational ethics, both normative and empirical scholars, shape future neuroscience research on topics germane to their research interests? Decision neuroscience confirms that people do not make ethical decisions in a strictly rational manner but that emotion and intuition play important roles. The challenge is to link neuroscience findings

D. C. Robertson (✉)
Legal Studies and Business Ethics Department, The Wharton School, University of Pennsylvania, Philadelphia, PA, USA
e-mail: robertsd@wharton.upenn.edu

J. T. Martineau, E. Racine (eds.), *Organizational Neuroethics*, Advances in Neuroethics, https://doi.org/10.1007/978-3-030-27177-0_9

such as these to questions specific to empirical organizational ethics research, as well as to higher-order normative questions. Research on moral cognition is emphasized, including conceptualizations of moral cognition, brain structure and function, and dual process theory and the role of emotion. Future research suggestions are included.

9.1 Introduction

Decision neuroscience is both a relatively new and a rapidly growing field, signifying that this is an opportune time to delineate its influence on organizational ethics, as well as to advocate in reverse for the influence of organizational ethicists to shape future neuroscience research. The ultimate goal (and we are far from reaching that goal) is to incorporate empirical findings from decision neuroscience into a theory of organizational ethics. Such a theory would lead not only to enhanced understanding of decision-making and behavior in the workplace but also would be the forerunner to developing means to intervene to improve those processes. Of course, this leads us inevitably to the normative question of what is meant by "improve," especially in the realm of ethical decision-making and behavior. We need to first determine what the desired ethical behavior is before we can devise interventions to achieve that behavior.

For purposes of this chapter, organizational ethics is a comprehensive term encompassing business ethics, as well as workplace ethics. (Note that much of the scholarship in organizational ethics has focused on business ethics, so that business ethics, while a narrower field of study than organizational ethics, arguably has a more robust body of scholarly literature.) Of relevance for organizational ethics, neuroscience research on moral cognition has increased dramatically over the last two decades and has already made significant contributions which relate to the field of ethics[1] [see, e.g., Greene and Cohen (2004), Shenhav and Greene (2014), and Young and Koenigs (2007)]. However, for the most part, this research does not connect to organizations and the workplace and does not incorporate an organizational ethics perspective. Further research that integrates organizational ethics and neuroscience is needed (Robertson et al. 2017; Salvador and Folger 2009).

The aim of decision neuroscience is to achieve a greater understanding of the biological mechanisms that mediate choice and decision-making and ultimately help to explain behavior (Butler et al. 2016). Organizational neuroscience is meant to complement organizational theories of attitudes and behaviors in the workplace by considering "the most fundamental level of analysis," "the basic building blocks of behavior," that is, brain function and activity (Becker et al. 2011). Decision neuroscience relies heavily, but not exclusively, on functional magnetic resonance

[1] Note that in this paper I use the terms "moral" and "ethical" interchangeably.

imaging (fMRI) technology, or brain scans. As such, decision neuroscience research holds considerable possibilities for organizational ethics scholars and opportunities for cross-disciplinary research. At the same time, we know that neuroscience has well-documented methodological limitations, but these are not the focus here.

It is difficult to discuss the empirical side of decision neuroscience and organizational ethics without also addressing normative questions. These normative questions are hardly new to organizational ethicists. Over more than the past two decades, scholars in the field of business ethics have observed the challenge of bridging the gap between the work of normative philosophers and empirical social scientists—with perhaps only modest success (Donaldson 1994; Robertson 1993; Treviño and Weaver 1994; Weaver and Treviño 1994). Conceivably the overall conclusion is that each stream of research can clearly stand on its own but that the field of business ethics is enriched when Weaver and Treviño's (1994) "symbiosis" occurs. In this symbiotic view, normative and empirical business ethics theory and methodology remain separate but rely on each other for the study of unethical behavior. This means that empirical inquiry must recognize the normative basis of the behavior in question. Of course, at the time that these scholars were writing in the 1990s, there was little to no awareness of neuroscience as a field that might have an impact on business ethics.

Now we are faced with the demanding and intriguing challenge to understand the role of cognitive neuroscience in both normative and empirical ethics research domains. Salvador and Folger (2009) believe that it is not just a question of integrating cognitive neuroscience and business ethics but that neuroscience holds promise for bridging the "normative-empirical divide" in that it necessitates greater conceptual clarity on the part of business ethics scholars. The question today becomes whether any symbiosis among neuroscientists, philosophers, and social scientists studying organizational ethics can be achieved.

This chapter addresses the following two research questions:

1. What are the most promising ways to enhance the field of organizational ethics through scholarly communication and collaborative work among philosophers, social scientists, and neuroscientists?
2. What contributions already have been made, and how can we build on those contributions? Specifically, how can scholars in the field of organizational ethics, both normative and empirical scholars, shape future decision neuroscience research on topics germane to their research interests?

The chapter will proceed as follows: Sect. 9.2 addresses how neuroscience plays a role in empirical organizational ethics research and how empirical organizational ethicists can direct future neuroscience research. This section emphasizes research on moral cognition and is the main focus of the chapter. Section 9.3 briefly focuses on the contributions of neuroscience to normative organizational ethics, as well as suggestions for how normative organizational ethicists can contribute to future neuroscience research. Section 9.4 addresses challenges and limitations associated with neuroscience. Section 9.5 concludes with thoughts about the role of organizational ethicists in collaborating with neuroscientists.

9.2 Neuroscience and Empirical Organizational Ethics

The orientation to date among organizational ethics researchers has been to explore ways in which cognitive neuroscience can contribute to organizational ethics, but I also examine the opposite point of view, asking how organizational ethicists can shape neuroscience research. This chapter identifies what I consider to be the most important and well-documented area in which neuroscience research can and does contribute to organizational ethics, the study of moral cognition [see, e.g., Casebeer and Churchland (2003), Hannah and Waldman (2015), and Moll et al. (2001, 2002)].

Cognitive neuroscience consists of a range of methodologies, including eye tracking (method in which a video camera records visual attention), electroencephalography (EEG) (method that places electrodes on the scalp to measure electromagnetic activity of the brain), transcranial magnetic stimulation (TMS), and transcranial direct current stimulation (tDCS) (methods that temporarily inhibit or stimulate specific brain areas or functions). The methodology most widely used in the study of moral cognition is that of fMRI technology. Participants are placed in an fMRI scanner and react to stimuli or perform tasks while their brain activity is measured. The most prominent measure is the blood-oxygen-level-dependency (BOLD) effect. BOLD fMRI captures changes in the oxygen concentration of the blood flow in the brain. Neural activity increases blood oxygenation. However, caution is needed as alternative explanations for the source of the blood oxygenation have been advanced (Logothetis et al. 2001).

9.2.1 What Do Cognitive Neuroscientists Mean by Moral Cognition?

This section explores how moral cognition is understood and studied in neuroscience research. The focus is to understand which areas of moral cognition are most and least studied by neuroscientists. The assumption is that neuroscientists have an explicit conceptualization of what constitutes moral cognition. As we will see, however, different conceptualizations of moral cognition are found in neuroscience research. Research on moral cognition investigates topics related to ethics but rarely related to the specific study of organizational ethics. The challenge for organizational ethics scholars is to devise ways to apply moral cognition neuroscience to a work setting.

A fundamental question is whether moral cognition means moral reasoning or sensitivity to moral issues or both. Rest (1984) proposed a four-component approach to moral behavior: moral sensitivity (awareness), moral judgment (or moral reasoning), moral motivation, and moral character. Rest emphasized the significance of moral sensitivity as a first step in his model. In the realm of organizational ethics, it is particularly important for individuals to recognize the ethical nature of an issue as part of their decision-making. As we will see, however, in the following examples of the neuroscience literature on moral cognition, recognition of the role of moral sensitivity is largely absent.

In addition to the lack of moral cognition research on moral sensitivity, two other conclusions relating to organizational ethics can be drawn. The first conclusion is that nearly all the neuroscience studies of moral cognition focus on the morality of causing harm to individuals, to the exclusion of other considerations of morality. The second conclusion is that very little neuroscience research on moral cognition considers ethical issues in the workplace. Considerable translation and creative research design are needed to apply findings from neuroscience studies of moral cognition to organizational ethics.

If we look to the field of moral psychology for conceptions of morality, we can then review examples of moral cognition neuroscience research with an eye toward the predominant understanding of morality. Graham et al. (2011) propose five universal foundations of morality: harm/care, fairness/reciprocity, in-group/loyalty, authority/respect, and purity/sanctity. In a later iteration, Graham et al. (2013) acknowledge that other elements could be added, most notably, liberty/oppression, suggested by Haidt (2012) as a sixth foundation. Of course, this categorization is not the only one developed by moral psychologists, but it is useful to apply it to neuroscience studies of moral cognition. The authors develop a Moral Foundations scale based on these six categories. The scale asks, "When you decide if something is right or wrong, to what extent are the following considerations relevant to your thinking?"

- Harm/Care: Whether or not someone suffered emotionally, for example, "Compassion for those who are suffering is the most crucial virtue."
- Fairness/Reciprocity: Whether or not some people were treated differently than others, for example, "Justice is the most important requirement for a society."
- In-group/Loyalty: Whether or not someone's actions showed love for his or her relevant group/community, for example, "People should be loyal to their family members, even when they have done something wrong."
- Authority/Respect: Whether or not someone showed a lack of respect for authority, for example, "Respect for authority is something all children need to learn."
- Purity/Sanctity: Whether or not someone violated standards of purity and decency, for example, "People should not do things that are disgusting, even if no one is harmed."

What is of interest about this list is that that most neuroscience studies tend to fall under the broad categories of harm/care to the exclusion of the other categories. No doubt the most famous and most dramatic neuroscience study of moral cognition is that of Greene et al. (2001) and the trolley dilemma. The trolley dilemma involves the decision to throw a switch to divert a runaway trolley from a track on which five people will be killed to a track on which one person will be killed instead. Greene also tested the variant of the trolley dilemma in which an individual must push a large stranger off a bridge, who will then be killed by the trolley, but will keep the trolley from killing the five people on the track. Greene studied brain activity of subjects as they encountered these two versions of the dilemma, concluding that pushing the large stranger evoked a great deal more emotion than did flipping the switch. But what is of note for present purposes is how this dilemma relates to moral cognition, particularly the classification scheme of Graham et al. (2011). Clearly it involves harm as individuals will be killed and the dilemma arguably ignores the other categories identified (fairness, loyalty, etc.). Similarly, Hauser et al. (2007), Young et al. (2010), and Young and Saxe (2008) focus on scenarios involving harm; in fact, most of Hauser's scenarios are variations of the trolley dilemma.

Young and Koenigs (2007) classify a set of ethical issues as addressing the question, "When, if ever, is it morally permissible to harm one to save many?" (p. 70). Indeed, there are multiple variations of this theme that neuroscientists have investigated. For example:

> *Transplant*: You are a doctor. You have five patients, each of whom is about to die due to a failing organ of some kind. You have another patient who is healthy.
>
> The only way that you can save the lives of the first five patients is to transplant five of this young man's organs (against his will) into the bodies of the other five patients. If you do this, the young man will die, but the other five patients will live (Greene et al. 2008).

> *Crying Baby:* Enemy soldiers have taken over your village. They have orders to kill all remaining civilians. You and some of your townspeople have sought refuge in the cellar of a large house. Outside you hear the voices of soldiers who have come to search the house for valuables.
>
> Your baby begins to cry loudly. You cover his mouth to block the sound. If you remove your hand from his mouth his crying will summon the attention of the soldiers who will kill you, your child, and the others hiding in the cellar. To save yourself and the others you must smother your child to death (Greene et al. 2008).

One must ask if harm should be the dominant focus of moral cognition in studies of organizational ethics, as this brief review suggests has been the case. Three of the categories of Graham et al. (2011) clearly also hold importance for organizations: fairness, loyalty, and authority, yet these are not receiving nearly as much neuroscientific scholarly attention.

A further conclusion is that for the most part neuroscience research on moral cognition tends to be abstract and devoid of the importance of task or setting. This is a shortcoming to be addressed in thinking about neuroscience findings and organizational ethics. This point can be illustrated with the work of Moll et al. (2001, 2006) and Mendez (2006). Moll et al. (2002) broadly conceptualized morality as either the welfare of society or that of other individuals. Stimuli were comprised of moral violations including physical assaults, poor children abandoned in the streets, and war scenes. Moll et al. (2001) studied moral cognition by presenting "moral" and contrasting "factual" statements to subjects. Subjects were asked to evaluate the statements as "right" or "wrong." The moral statements included:

Old people are useless
They hung an innocent
Every human being has the right to live
The boy stole his mother's savings
The wounded were left behind

Similarly, Mendez (2006) presents subjects with a series of statements ranging from "Take the last seat on a crowded bus" to "Drive after having one drink" to "Drive out the homeless from your community." This list of statements does include two that relate directly to workplace ethics, but these are the exception in terms of acknowledging that moral issues can take place in a workplace setting. The two are, "Take credit for others' work" and "Get more time off than your co-workers."

A further exception is that of Greene et al. (2001) as the following examples from their work illustrate.

> You are a young architect visiting one of your construction sites with your boss. Your boss is a despicable individual who makes everyone around him miserable including you. It occurs to you that if you were to push him off of the building you are inspecting, he would fall to his death and everyone would think it was an accident. Is it appropriate for you to push your boss off the building in order to get him out of your life?

So yes, broadly construed this is a workplace issue, but murder is a legal issue, and this scenario does not fit any standard framework of organizational ethics issues. Similarly, the following are workplace issues, but again the vignettes involve illegal actions, rather than organizational ethics issues.

> You are a management consultant working on a case for a large corporate client. You have access to confidential information that would be very useful to investors. You have a friend who plays the stock market. You owe this friend a sizable sum of money.
>
> By providing her with certain confidential information you could help her make a lot of money, considerably more than you owe her. If you did this, she would insist on canceling your debt. Releasing information in this way is strictly forbidden by federal law.
>
> Is it appropriate for you to release this information to your friend so that she will cancel your debt?

Or

> You are the owner of a small business trying to make ends meet. It occurs to you that you could lower your taxes by pretending that some of your personal expenses are business expenses.
>
> For example, you could pretend that the stereo in your bedroom is being used in the lounge at the office, or that your dinners out with your wife are dinners with clients.
>
> Is it appropriate for you to pretend that certain personal expenses are business expenses in order to lower your taxes? (Greene et al. 2001).

It should also be noted that the Greene et al.'s (2001) studies have been criticized from a methodological standpoint (Kahane and Shackel 2010) but are being used here for purposes of illustrating the scarcity of business or organizational ethics vignettes in the neuroscience literature on moral cognition.

As these examples demonstrate, it is difficult to find neuroscience studies of moral cognition that relate to organizational ethics, that is, both that have a workplace or business context and that contain ethical rather than legal issues. The paucity of organizational ethics scenarios in neuroscience studies of moral cognition suggests an opportunity for organizational ethicists to develop and test scenarios that relate to moral decisions having to do with business and the workplace. Scenarios could be constructed that examine neural activity as individuals react to hiring, promotion, and firing incidents perceived to reflect unfair employment practices. Variables such as whether one is the victim or the observer of the unfair practices could be included. Social justice theory is fundamental to the study of organizational ethics yet has been largely ignored in neuroscience research. Additionally, neuroscience research could study neural activations to situations in which employees demonstrate insubordination to authority or believe that the organization is constraining their liberty. The opportunities for organizational decision neuroscience research are extensive.

Additionally, it would be instructive to develop an organizational ethics conception of moral cognition, which presumably would go beyond the notion of harm. Thus, those of us in the field of organizational ethics are presented with an opportunity to shape the nature of the study of moral cognition in future neuroscience studies. Perhaps we can reference Graham et al. (2013) and other conceptualizations of morality that go beyond issues of harm and care. Possible research questions around fairness and respect for authority in the workplace have been mentioned. Loyalty in the workplace is also a noteworthy topic. One can envision studies of brain activity as individuals react to stimuli depicting employees who display signs of loyalty or even more telling signs of disloyalty. Thus, we can begin a discussion of how to design neuroscience projects to investigate the research questions and agendas most fundamental to organizational ethicists.

9.2.2 Brain Structure and Function

Closely associated with conceptualizations of moral cognition is the intriguing question of whether and how moral decision-making may differ from other forms of decision-making. Of course, all cognitive decision-making has certain elements in common, but moral decision-making can involve different forms of neural activity. Perhaps this is best illustrated by the famous case of Phineas Gage, who sustained a brain injury, but for the most part maintained his cognitive reasoning ability. His behavior, however, suggests that he lacked the ability to make appropriate moral decisions (Eslinger and Damasio 1985). Subsequent research on neurologically impaired subjects has corroborated that subjects may retain their normal cognitive abilities but at the same time exhibit unethical behavior (Greene 2005).

Since the time of the Phineas Gage case, neuroscience research has begun to map moral judgment in the brain. Over time, moral cognition has been relatively consistently associated with several brain regions (anterior prefrontal cortex, orbitofrontal cortex, posterior superior temporal sulcus, anterior temporal lobes, insula, precuneus, anterior cingulate cortex, and the limbic system) (Moll et al. 2005). Although early studies such as the Phineas Gage case found areas specifically associated with moral cognition, more recent work acknowledges the challenges of doing so. The current conclusion is more complex, acknowledging that the connectivity of brain regions means that moral judgment and cognition are found throughout the brain (Jack et al. 2013; Pujol et al. 2011).

Note also that the regions that have been associated with moral cognition are also associated with other reasoning and decision-making processes, making it more difficult to isolate brain regions specific to moral cognition. Activation of brain regions differs according to the situational cues of a particular study design and is dependent on the nature of the moral problem, the difficulty of the decision, and contextual stimuli. Robertson et al. (2007) found activations of areas of the brain associated with ethical decision-making that were distinct from areas associated with business strategic decision-making. This finding suggests that future research could investigate the circumstances under which brain activity associated with ethical decision-making can be isolated from other types of decision-making.

A recent study attempts to corroborate brain structure features underlying Kohlberg's (1958) classic theory of moral reasoning (Prehn et al. 2015). Kohlberg's cognitive-developmental approach dominated moral psychology for decades. Kohlberg posited that individuals pass through specific stages of moral reasoning in an invariant sequence. Once a particular stage is reached, it is not possible to go back to the previous stage. This theory has provided the foundation for significant research delineating the thought processes of individuals at the various stages. However, there is nothing in the theory that provides any physiological evidence that these stages exist. The research by Prehn et al. (2015) found that the brains of subjects at the post-conventional level of moral reasoning were characterized by increased gray matter volume in the ventromedial prefrontal cortex and subgenual anterior cingulate cortex, compared with subjects at a lower level of moral reasoning. This line of research holds promise for investigating the brain structure of individuals in relation to their ethical decision-making. In other words, do the brains of individuals who make an ethical decision differ from the brains of individuals who make unethical decisions, and, if so, how? Of course, knowing this will not answer the overarching question of cause and effect, that is, whether different brain structure precedes different moral reasoning, or vice versa.

One can conceptualize ways in which knowledge of different brain structure could be advantageous in the workplace. For example, many firms engage in ethics training without conclusive evidence as to its effect. If neuroscience studies could be designed to show that such training can result in differing brain structure or function, either immediately or over time, this would provide a physiological marker as to their potential effectiveness. Neuroimaging technology would need to advance considerably to achieve conclusive results about the effects of ethics training [see, e.g., Illes and Racine (2005)]. Of course, firms are also trying to measure the effectiveness of ethics training both attitudinally and behaviorally. With advances in technology, future neuroscience results potentially could complement psychometric data in the complex process of measuring the results of ethics training.

9.2.3 Dual Process Theory and the Role of Emotion

According to dual system or dual process theory, individuals make decisions using both intuition and reasoning (Cushman 2013; Greene 2014). The first process is one that is automatic and favors what is happening now, whereas the second process is what Greene terms "manual" and involves a more reasoned approach favoring what will be happening later. Greene demonstrates through his own research and that of others that distinct areas of the brain are involved in each of these two processes. Haidt (2001) has been particularly convincing about the significance of the automatic process, emphasizing the role of intuition in ethical or moral decision-making. Building on this dual process model, Reynolds (2006a) proposes a "neurocognitive model of ethical decision making." His model specifies how individuals think as they make decisions about ethical issues using both reflexive unconscious reactions and a higher-order or reasoning process of thinking. Reynolds believes that a great deal of reaction to ethical issues occurs below the level of consciousness and is responsive to an individual's own ethics template that is both descriptive and normative. Reynolds (2006b) also finds that a manager's ethical predispositions influence their responses to the very nature of the moral issue, further complicating the decision-making process.

In other words, an individual decision-maker faced with the decision whether to cheat on their expense account will have an unconscious sense of the descriptive element of what is meant by cheating on an expense account cheating (e.g., large or little amount and organizational norms for overlooking or even condoning small amounts), as well as ethical predispositions informing the normative basis of why it is wrong to do so. However, it would be remiss not to acknowledge the methodological controversies involved in this research. Dubljević and Racine (2014) provide an excellent critical review of this literature. One of the many points they make stands out in relation to the study of organizational ethics: these studies tend to confront individuals with highly unrealistic moral dilemmas.

Neuroscience technology seems particularly appropriate for the investigation of the role of automatic processes in ethical decision-making, although advances in temporal resolution are needed to capture the rapid nature of these processes. An attractive feature of fMRI technology is that it eliminates the problem of social desirability bias so common in the study of business ethics. For the most part, individuals believe themselves to be more ethical than other people (Messick and Bazerman 2001), and they also want to give the appearance of being more ethical (Randall and Fernandes 1991). If a survey question asks individuals if they have cheated on their expense accounts, for example, it may be assumed that the most unethical people are also the most likely to lie and tell you that they are ethical, in order to present themselves in the most favorable possible light. But individuals cannot control neural activity in their brain, and the idea is that fMRI data will represent accurately the attitudes of an individual, regardless of their social desirability. As such, fMRI technology suggests significant potential but is also subject to limitations [see, e.g., Wolpe et al. (2005)]. The thinking here is similar to the rationale for the administration of a lie detector test, or polygraph. The usefulness of the polygraph lies in its ability to tap into physiological indices of truth telling and lying that are not under the control of the individual. Similarly, fMRI studies have explored neural responses as individuals deceive others [see, e.g., Abe et al. (2007) and Farah et al. (2014)].

Neuroscience studies could be designed to determine the circumstances under which behavior in an organization, such as lying, is perceived to be acceptable. Such studies would not be confined to lying but could include any number of potential instances of misbehavior in the workplace. In other words, if asked in a survey, employees may answer that it is always wrong to lie to your boss or to your fellow employees, but examining what is going on in the brain when subjects see stimuli describing such behavior demonstrates that they actually condone lying in these circumstances. Similar studies could investigate, for example, brain activity as a reaction to depictions of sexual harassment, unfair promotion practices, or discrimination in the workplace.

With the emergence of the relatively new field of behavioral ethics and its emphasis on intuition and decisions that occur below the level of consciousness, neuroscience has a great deal to contribute. Of course, it is important to be only cautiously optimistic about the promise of neuroscience; one should not overlook the significant methodological limitations that have been raised [see, e.g., Racine et al. (2017) for a review].

Broadly defined, behavioral ethics is "concerned with explaining individual behavior that occurs in the context of larger social prescriptions" (Treviño et al. 2006, p. 952). In particular, behavioral ethics seeks to explain the curious and frequently observed phenomenon that good people do bad things. Individuals seem able to engage in a range of ways to deny that their actions are unethical, many of them occurring below the level of consciousness, including simply not recognizing the ethical dimension of a decision, as well as rationalizations as to why the action is ethical. Moral neglect occurs when individuals fail to see or overlook the ethical dimension of their actions (Moore and Gino 2013; Tenbrunsel and Messick 2004). Ethical fading occurs when decision-makers falsely believe that they are upholding their moral principles, but in fact, the ethical aspects of the decision fade into the background as more compelling and self-interested aspects of the decision obscure the importance of ethical considerations (Tenbrunsel and Messick 2004). Unconscious biases or blind spots also account for unethical behavior, and cognitive neuroscience is well positioned to study them (Bazerman and Tenbrunsel 2011). For example, one can imagine an eye-tracking study in which individuals respond to scenarios depicting protagonists of different races. Eye tracking could be used to determine what individuals literally see and don't see in such scenarios, exposing the very nature of blind spots.

Organizational contexts can be structured with prompts that will lead individuals to make specific ethical decisions (Bowie 2009), in other words, a "nudge." Thaler and Sunstein (2009) acknowledge that there can be "evil nudges" and "bad nudges." This means that nudges can be put in place in organizations to benefit the employer at the expense of the employee. However, in the case of nudges that lead to employee ethical behavior, the assumption is that both employer and employee benefit. Take a simple example such as display of "moral symbols" in organizations. Desai and Kouchaki (2017) demonstrate that exposure to such symbols presented by subordinates in an organization can dissuade their superiors from engaging in unethical behavior, as well as lower the likelihood that the subordinates will be asked to engage in unethical behavior.

It would be interesting to design neuroscience studies that track brain activity as individuals react to specific moral symbols. Subjects could be exposed to particular symbols, for example, and then asked to make a decision involving an ethical issue. The design of the study would involve investigating the impact of the symbols on the decisions made but would go further in exploring brain activity both as individuals are exposed to the symbols and as they make their decisions.

A further topic of study relevant to the workplace is the role of compensation and reward systems. Perhaps it can be argued that compensation systems in and of themselves constitute a less than subtle nudge for employees to act in certain ways. There is also preliminary evidence that acting ethically can provide its own form of intrinsic reward. Moll et al. (2006) found that areas of the brain connected to reward, specifically the mesolimbic reward system, are activated to a similar extent when individuals engage in philanthropy as when they receive financial rewards. Similarly, Fang et al. (2017) found that individuals exhibiting higher stages of moral reasoning also had increased activation in the mesolimbic frontostriatal system which plays an important role in reward processing. Cognitive neuroscience can advance our understanding of how employees in the workplace process various forms of reward, including monetary rewards, recognition, and rewards stemming from cooperative or altruistic behavior, and how these in turn influence ethical decision-making and behavior. For example, there may be differences in brain activity associated with various kinds of rewards, such as bonuses or pay raises, immediate versus longer-term rewards, or recognition versus monetary rewards.

Neuroscience research in moral cognition has demonstrated the importance of the role of emotion in ethical decision-making [see, e.g., Moll et al. (2002) and Young and Koenigs (2007)]. Research has shown that the prefrontal cortex and also the regions of the brain involved in emotional reactions are important for a person's moral development (Greene 2015). Combined with the importance of unconscious processes, this emphasis on emotion validates that decision-making about ethical issues is not solely a conscious decision by rational actors. A rational decision-maker faced with the question of whether or not to behave unethically would be expected to consider the likelihood of detection and punishment, the burden imposed by that punishment, and the attractiveness of the potential reward associated with acting unethically. Instead neuroscience establishes that such calculation is at best only one aspect of how individuals think as they engage in decision-making about an ethical issue (Cushman 2013).

9.2.4 Further Neuroscience Topics for Organizational Ethics

Many possibilities exist for future decision neuroscience research applicable to organizational ethics. In the following sections, I identify three areas that I believe are especially relevant.

Mirror Neurons In the workplace, mirror neurons operate so that workers automatically and often unconsciously imitate the attitudes and behaviors of others (Becker and Cropanzano 2010). This phenomenon is especially important to the study of organizational ethics. We know that individuals in the workplace respond to the actions of referent others in their ethical decision-making and behavior [see, e.g., Treviño (1986) and Schwartz (2016)]. Mirror neurons provide an explanation for why this would be the case. Mirror neurons are a type of brain cell that responds equally when an individual performs an action and when an individual witnesses someone else perform the same action. In the world of teenagers, peer pressure explains a great deal of seemingly foolish or irrational behavior (Albert et al. 2013). There can be a similar contagion effect of both ethical and unethical behaviors in organizations. One caution to be exercised and studied is the potentially negative role of in-group bias on the effects of mirror neurons. In other words, mirror neurons may only come into play with members of one's own inner circle. Further research is needed to explore exactly when and how mirror neurons operate in organizations. For example, research could investigate whether mirror neurons only are activated between coworkers who hold similar positions in an organization, or who work together on a regular basis, or who are similar in demographic characteristics.

Empathy Mirror neurons are also believed to be related to empathy in organizations, which in turn is associated with prosocial behavior (Corradini and Antonietti 2013; Decety 2014). However, we may need to exercise caution and engage in further research to be able to draw conclusions that verify these associations (Lamm and Majdandžić 2015). Empathy in organizations allows employees to function well on teams and to exhibit care for their peers. As with mirror neurons, a further caution is needed, however, in that empathy may not always extend to fellow employees who are on different teams or who belong to different ethnic groups. Empathy may not extend to those who are perceived as "other."

Trust and Cooperation Cognitive neuroscience studies have identified brain regions associated with both trust and cooperation, particularly using trust games (Johnson and Mislin 2011; Krueger et al. 2007). Also of interest is the study of Dimoka (2010) that establishes distinct brain activity associated with trust and distrust. This finding suggests that the study of organizational ethics could pursue both lines of inquiry, that is, the study of the effects of trust in the organization, as well as the effects of distrust. If trust is the foundation of ethical workplace behavior, it seems plausible to hypothesize that distrust leads to lack of workplace cooperation and to unethical behavior. Neuroscience research could be designed to test how individuals react to perceptions of betrayal of trust on the part of work colleagues and the impact this has on projected behavior.

9.3 Neuroscience and Normative Business Ethics

It is perhaps easier to argue that normative business ethics has a great deal to communicate and contribute to empirical business ethics than it is to make the opposite claim about the contribution of empirical business ethics to normative business ethics. There is a sense that empirical work must always acknowledge normative work, whereas normative work need not always acknowledge empirical influence (Robertson 1993). Let's take, for example, the hypothesis that corporate codes of ethics will lead to less unethical behavior on the part of employees. This hypothesis makes no sense without normative underpinnings as to what constitutes employee unethical behavior. Recently, using the concept of corporate governance, Donaldson (2012) argued convincingly for the importance of making the normative basis of corporate practice explicit. He contends that much of agency theory, a self-described positive theory of economic behavior, rests on normative assumptions, including that agents are justified to act in the best interests of their principals, in other words, that this is a good thing. As Donaldson demonstrates, the normative basis of positive economic theories of the firm is often ignored.

This is indeed the challenge. Researchers trained in a specific discipline tend to ignore the potential contribution of their counterparts in other disciplines. Normative work may depend on context, and empirical work may rest on normative foundations, but these influences, where they exist, have tended to be implicit. This implicit nature is especially apparent in neuroscience research. Various neuroscience studies of moral cognition, for example, have used multiple conceptions of what is meant by moral cognition without making explicit exactly what comprises moral cognition, the circumstances in which it occurs, and how it leads to ethical or unethical behavior. Nor have neuroscientists tended to explore or make explicit the normative basis of moral cognition. For the most part, vignettes or statements used in neuroscience research are labeled "moral" without an accompanying explanation as to what makes them moral.

An exception is the contention of Greene (2014) that neuroscience does matter for our understanding of normative ethics. Furthermore, and this is of interest to normative business ethics research, Greene attempts to establish that deontological thinking is automatic, whereas utilitarian thinking is manual. These results have been contested, but Greene makes the sweeping claim that the neurological basis of ethical thinking will lead us to revise our normative ethics theories. Greene also argues "that a deeper understanding of moral psychology favors certain forms of consequentialism over other classes of normative moral theory" (p. 695). Here is his argument:

> Science can advance ethics by revealing the hidden inner workings of our moral judgments, especially the ones we make intuitively. Once those inner workings are revealed we may have less confidence in some of our judgments and the ethical theories that are (explicitly or implicitly) based on them (pp. 695–696).

Thus, Greene's contention is that the impact of cognitive neuroscience on normative ethics rests on an understanding of empirical neuroscience research on decision-making processes and moral judgments.

Bowie (2009) studied the specific topic of the impact of brain research, as well as psychological research, on philosophical ethics, concluding that there are conditions under which human freedom and responsibility for choice may be diminished by brain impairment. However, as Bowie (2009) rightfully points out, knowing which part of the brain houses moral cognition or how individuals engage in moral reasoning does not tell us how to resolve a moral problem, nor does it help us to know the essence of the meaning of moral cognition or whether an action is justified ethically. Bowie also argues that in some cases, knowledge about brain activity may seem to restrict our freedom of choice, but in other instances, such knowledge will enhance understanding of freedom and responsibility if we know how to act on the information provided by neuroscience. Although Bowie does not elaborate exactly how this can be accomplished, one can imagine, for example, that awareness of and knowledge about our unconscious biases could lead us to modify our behavior to overcome those biases.

Even if, as Churchland (2013) asserts, cognitive neuroscience cannot by itself provide explanations for attitudes or behavior, it can document their source in a manner that may be fruitful for normative research. In addition to the concepts of consciousness and moral identity, normative business ethics is concerned with issues of free will, autonomy, agency, intent, and responsibility. Cognitive neuroscience can tell us how brain damage or impairment affects each of these phenomena. Recent work at the intersection of law and neuroscience has even found detectable differences in the brain between individuals who are knowing versus those who are reckless as they contemplate a crime (Vilares et al. 2017). These categories are central to criminal convictions and sentencing but often prove elusive. Similarly, the notion of intent is focal to studies of organizational ethics. If neuroscience can reveal an individual's intent, deliberate or accidental, this could potentially change our attributions of an individual's moral responsibility.

All this points to the fact that a great deal of thought and research looks at the question of what cognitive neuroscience may contribute to the field of normative business ethics. But what if we turn this around and ask a different question, that is, what does normative organizational ethics have to contribute to cognitive neuroscience? My answer would be that there is a great deal to learn and a great deal of further research needed to understand the normative underpinnings of cognitive neuroscience and its study of moral cognition. Some basic questions remain. Are cognitive neuroscience studies using the right conceptualizations of the moral brain? Could normative organizational ethicists propose ethical issues to be studied beyond obvious examples, such as lying? What are the normative areas that have been neglected by cognitive neuroscience, and how can these gaps be filled? As in the previous discussion of empirical business ethics research, perhaps it would be useful for normative business ethicists to formulate and present their own conceptualization of moral cognition for research by cognitive neuroscientists.

9.4 Challenges and Limitations

Cognitive neuroscience requires unique positioning in relation to organizational ethics research. This differentiation hinges on at least two critical elements:

1. Cognitive neuroscience is a field of study concerned with the biological processes and physical aspects of the brain underlying cognition, including mental processes, experiences, sensations, intuitions, and understandings. The power of this methodology lies in its ability to investigate in real time what is happening in the brain as individuals make decisions. Cognitive neuroscience is able to tell us what part of the brain is activated in response to a particular stimulus, but research demonstrating why that part of the brain is activated is still in a relatively exploratory phase. Unlike other social science research in business ethics, cognitive neuroscience, at least at its present stage of development, is more descriptive than explanatory.
2. There is a greater translation and accessibility challenge with neuroscience findings than with other empirical findings in business ethics. Neuroscience requires specialized knowledge and uses specific methodologies not found in other social sciences. Its influence on organizational ethics depends largely on an ability to understand these methodologies and their accompanying specialized language and algorithms. Also, it is important to note the criticisms of neuroscience findings, perhaps especially that brain scan interpretations using reverse inference overclaim their application to decision-making and behavior.

9.4.1 Neuroscience as Descriptive

As a descriptive methodology, fMRI findings provide correlations with a psychological state that is associated with some level of heightened neuron activity. The resulting changes in the magnetic field measured by fMRI are used as a proxy for brain areas that are active during a task or as a response to a stimulus (Dimoka 2012; Huettel et al. 2004). For example, the mesolimbic reward system is activated when individuals make donations to charitable organizations in the same way as when individuals receive monetary rewards (Moll et al. 2006). However, Churchland (2013) argued that these correlations do not necessarily comprise explanations of attitudes or behavior. Indeed, criticism of neuroscience findings contends that there is a tendency toward overinterpretation of results. The assertion is that explanations of states of mind are extrapolated from descriptions of neural activity and that correlation may be confused with causation (Rachul and Zarzeczny 2012). Despite Churchland's contention that fMRI findings do not provide explanations, Kable (2011) maintains that causal inferences can be made in the sense that changes in the environment or in psychological states cause changes in brain activity. This set of arguments suggests that we should continue to seek ways in which neuroscience can be helpful in explaining ethical decision-making and behavior, but at the same time, we should exercise caution in attributing explanatory powers to neuroscience findings.

9.4.2 Accessibility of Neuroscience Findings

The challenges of working across scholarly disciplines are many. One problem of integrating the work of normative and empirical researchers is communication. Treviño and Weaver (1994) illustrate this problem with a (presumably) hypothetical conversation between a philosopher and a business school faculty member at an academic conference. In this conversation, each scholar denigrates the research approach of the other. To follow Treviño and Weaver's (1994) lead, one can only imagine what a hypothetical conversation would look like among neuroscientists, philosophers, and social scientists in organizational ethics. It could go something like this:

> Neuroscientist: My research has found that a decision-making scenario about ethical issues elicits emotion.
>
> Social scientist organizational ethicist: We already know that. We don't need brain science to tell us what we already know. And just because there is neural activity in the part of the brain housing emotion, how can you be sure that emotion is tied directly to ethics? Maybe something else in the scenario is triggering emotion. Besides, I read your paper and I could not make heads or tails out of your statistical analysis. It is as if you are speaking a different language.
>
> Philosopher organizational ethicist: How does it help me to know that emotion is involved in decision making about an ethical issue? That doesn't tell me anything about whether the decision is a good one.

Neuroscience methodologies are often dissimilar enough from other organizational ethics methodologies that neuroscience research results may remain opaque to those not trained in the methodologies. Perhaps it is incumbent on neuroscientists to communicate their research questions and findings in a manner that is comprehensible to organizational ethics scholars not schooled in neuroscience technology. Despite this need for specialized training, neuroscience findings have found their way into the popular press, and these findings appear to have an appeal to a general audience. Weisberg (2008) found that people are unduly impressed by neuroscience findings perhaps because they are accompanied by visually appealing brain pictures. Thus, the media can translate neuroscience findings for the lay public but in a superficial way. There is a further need for more rigorous translation into terms that are relevant, actionable, and understood by scholars in other disciplines. Neuroscientists also need to be forthcoming in acknowledging the limitations of their methodology.

One approach is provided by Kable (2011) who offers a toolkit of the methods of cognitive neuroscience meant for an audience that has "very little, if any, knowledge of cognitive science techniques" (p. 63). Kable outlines three types of tests common to fMRI methodology: tests of association, tests of necessity, and tests of sufficiency. Tests of association "involve observing or experimentally manipulating psychological states of behavior, simultaneously measuring neural activity, and examining the correlation between the two" (p. 65). For example, studies of cooperation show consistent activation in brain areas that have been associated with reward processing [see, e.g., Rilling et al. (2002)]. Tests of necessity "involve disrupting neural activity and showing that this manipulation impairs a specific behavioral or psychological function" (p. 66). Finally, tests of sufficiency "involve enhancing neural activity and showing that this manipulation results in a specific behavioral or psychological state" (p. 66). Kable's tests of association tend to be the most relevant for organizational ethics researchers, as well as perhaps the most easily understood by non-neuroscience scholars. Nevertheless, the presentation of neuroscience results is not always accessible to those not trained in neuroscience.

9.5 Conclusion

The aim of this chapter has been to begin a conversation about how those of us in the field of organizational ethics can learn from neuroscience, as well as how we can contribute to neuroscience research. As discussed, the challenges are many, but the opportunities are pronounced. It is incumbent upon organizational ethicists at least to familiarize ourselves with the methodology of neuroscience research, its analysis, and interpretation. We need to be able to read articles with some level of knowledge and sophistication and to have a sense of which research questions in our field are compatible with neuroscience technology. The field of organizational ethics has much to learn from neuroscience research, as well as much to teach it.

This chapter has contended that neuroscience is different from other empirical social sciences and thus perhaps more challenging to integrate with our field of organizational ethics; nevertheless, we would be remiss in ignoring its potential ability to contribute to understanding of ethical decision-making and behavior. Noteworthy findings in moral cognition already hold promise for shaping future organizational ethics research. As this chapter has discussed, we can identify neural activity underlying ethical decision-making and, to a certain extent, can differentiate it from neural activity associated with other types of decision-making. We know that emotion and intuition play a sizable role in ethical decision-making. We also know that impairment in certain areas of the brain can alter ethical decision-making, as well as ethical behavior. Clearly there is much that we still don't know and that, as organizational ethicists, we would like to know. We have questions that neuroscientists can help to answer.

What does this look like in practice? Organizational ethicists can work with neuroscientists to design interdisciplinary research to address foremost questions in the field, whether it be how to influence unconscious biases, how to structure organizational rewards that encourage employee ethical behavior, or how choice architecture can work to elicit ethical behavior. We would like to know, for example, what goes on in the brain as individuals face questions about how to prioritize stakeholder interests. We know that trust and distrust look different in the brain (Dimoka 2010). Perhaps cooperation and competition do as well. Additionally, and perhaps more importantly, as organizational ethicists, we can advance our own conceptualization of moral cognition, one that goes beyond scenarios depicting harm. We would like to know more about the brain's neural activity underlying concepts of justice, duty, and rights, to name only a few seminal theories in business ethics.

As this chapter has noted, normative and empirical works in organizational ethics benefit when each stream acknowledges the considerable role of the other. Similarly, it would be a missed opportunity if organizational ethicists treat neuroscience as a completely separate field of study. Neuroscience is still in a relatively nascent stage, which affords the prospect of influencing future work to further understand how the brain processes ethical issues in the workplace.

References

Abe N, Suzuki M, Mori E, Itoh M, Fujii T. Deceiving others: distinct neural responses of the prefrontal cortex and amygdala in simple fabrication and deception with social interactions. J Cogn Neurosci. 2007;19(2):287–95.

Albert D, Chein J, Steinberg L. Peer influences on adolescent decision-making. Curr Dir Psychol Sci. 2013;22:80–6.

Bazerman MH, Tenbrunsel AE. Blind spots: why we fail to do what's right and what to do about it. Princeton: Princeton University Press; 2011.

Becker WJ, Cropanzano R. Organizational neuroscience: The promise and prospects of an emerging discipline. J Organ Behav Manage. 2010;31(7):1055–9.

Becker WJ, Cropanzano R, Sanfey AG. Organizational neuroscience: taking organizational theory inside the neural black box. J Manag. 2011;37(4):933–61.

Bowie NE. How empirical research in human cognition does and does not affect philosophical ethics. J Bus Ethics. 2009;88(4):635–43.

Butler MJ, O'Broin HL, Lee N, Senior C. How organizational cognitive neuroscience can deepen understanding of managerial decision-making: a review of the recent literature and future directions. Int J Manag Rev. 2016;18(4):542–59.

Casebeer WD, Churchland PS. The neural mechanisms of moral cognition: a multiple-aspect approach to moral judgment and decision-making. Biol Philos. 2003;18(1):169–94.

Churchland PS. Touching a nerve: our brains, our selves. New York: WW Norton & Company; 2013.

Corradini A, Antonietti A. Mirror neurons and their function in cognitively understood empathy. Conscious Cogn. 2013;22(3):1152–61.

Cushman F. Action, outcome, and value: a dual-system framework for morality. Personal Soc Psychol Rev. 2013;17(3):273–92.

Decety J. The neuroevolution of empathy and caring for others: why it matters for morality. In: Decety J, Christen Y, editors. New frontiers in social neuroscience. New York: Springer; 2014. p. 127–51.

Desai SD, Kouchaki M. Moral symbols: a necklace of garlic against unethical requests. Acad Manag J. 2017;60(1):7–28.

Dimoka A. What does the brain tell us about trust and distrust? Evidence from a functional neuroimaging study. MIS Q. 2010;34:373–96.

Dimoka A. How to conduct a functional magnetic resonance (fMRI) study in social science research. MIS Q. 2012;36:811–40.

Donaldson T. When integration fails: the logic of prescription and description in business ethics. Bus Ethics Q. 1994;4(2):157–69.

Donaldson T. The epistemic fault line in corporate governance. Acad Manag Rev. 2012;37 (2):256–71.

Dubljević V, Racine E. The ADC of moral judgment: opening the black box of moral intuitions with heuristics about agents, deeds, and consequences. AJOB Neurosci. 2014;5(4):3–20.
Eslinger PJ, Damasio AR. Severe disturbance of higher cognition after bilateral frontal lobe ablation patient EVR. Neurology. 1985;35(12):1731–41.
Fang Z, Jung WH, Korczykowski M, Luo L, Prehn K, Xu S, Detre JA, Kable JW, Robertson DC, Rao H. Post-conventional moral reasoning is associated with increased ventral striatal activity at rest and during task. Sci Rep. 2017;7(1):7105.
Farah MJ, Hutchinson JB, Phelps EA, Wagner AD. Functional MRI-based lie detection: scientific and societal challenges. Nat Rev Neurosci. 2014;15(2):123.
Graham J, Nosek BA, Haidt J, Iyer R, Koleva S, Ditto PH. Mapping the moral domain. J Pers Soc Psychol. 2011;101(2):366.
Graham J, Haidt J, Koleva S, Motyl M, Iyer R, Wojcik SP, Ditto PH. Moral foundations theory: the pragmatic validity of moral pluralism. In: Devine P, Plant A, editors. Advances in experimental social psychology, vol. 47. San Diego: Academic Press; 2013. p. 55–130.
Greene J. Cognitive neuroscience and the structure of the moral mind. In: Carruthers P, Laurence S, Stich S, editors. The innate mind: structure and contents, vol. 1. New York, NY: Oxford University Press; 2005. p. 338–52.
Greene JD. Beyond point-and-shoot morality: why cognitive (neuro) science matters for ethics. Law Ethics Hum Rights. 2014;9(2):141–72.
Greene JD. The cognitive neuroscience of moral judgment and decision making. In: Decety J, Wheatley T, editors. The moral brain: a multidisciplinary perspective. Cambridge, MA: MIT Press; 2015. p. 197–220.
Greene J, Cohen J. For the law, neuroscience changes nothing and everything. Philos Trans R Soc B. 2004;359(1451):1775.
Greene JD, Sommerville RB, Nystrom LE, Darley JM, Cohen JD. An fMRI investigation of emotional engagement in moral judgment. Science. 2001;293(5537):2105–8.
Greene JD, Morelli SA, Lowenberg K, Nystrom LE, Cohen JD. Cognitive load selectively interferes with utilitarian moral judgment. Cognition. 2008;107(3):1144–54.
Haidt J. The emotional dog and its rational tail: a social intuitionist approach to moral judgment. Psychol Rev. 2001;108(4):814.
Haidt J. The righteous mind: why good people are divided by politics and religion. New York: Vintage; 2012.
Hannah ST, Waldman DA. Neuroscience of moral cognition and conation in organizations. In: Waldman DA, Balthazard PA, editors. Organizational neuroscience. Bingley: Emerald Group; 2015. p. 233–55.
Hauser M, Cushman F, Young L, Kang-Xing Jin R, Mikhail J. A dissociation between moral judgments and justifications. Mind Lang. 2007;22(1):1–21.
Huettel SA, Song AW, McCarthy G. Functional magnetic resonance imaging. Sunderland: Sinauer Associates; 2004.
Illes J, Racine E. Imaging or imagining? A neuroethics challenge informed by genetics. Am J Bioeth. 2005;5(2):5–18.
Jack AI, Dawson AJ, Norr ME. Seeing human: distinct and overlapping neural signatures associated with two forms of dehumanization. NeuroImage. 2013;79:313–28.
Johnson ND, Mislin AA. Trust games: a meta-analysis. J Econ Psychol. 2011;32(5):865–89.
Kable JW. The cognitive neuroscience toolkit for the neuroeconomist: a functional overview. J Neurosci Psychol Econ. 2011;4(2):63.
Kahane G, Shackel N. Methodological issues in the neuroscience of moral judgement. Mind Lang. 2010;25(5):561–82.
Kohlberg L. The development of modes of moral thinking and choice in the years 10 to 16. Chicago: University of Chicago; 1958.
Krueger F, McCabe K, Moll J, Kriegeskorte N, Zahn R, Strenziok M, Heinecke A, Grafman J. Neural correlates of trust. Proc Natl Acad Sci USA. 2007;104(50):20084–9.

Lamm C, Majdandžić J. The role of shared neural activations, mirror neurons, and morality in empathy—a critical comment. Neurosci Res. 2015;90:15–24.

Logothetis NK, Pauls J, Augath M, Trinath T, Oeltermann A. Neurophysiological investigation of the basis of the fMRI signal. Nature. 2001;412(6843):150.

Mendez MF. What frontotemporal dementia reveals about the neurobiological basis of morality. Med Hypotheses. 2006;67(2):411–8.

Messick DM, Bazerman MH. Ethical leadership and the psychology of decision making. In: Dienhart J, Moberg D, Duska R, editors. The next phase of business ethics: integrating psychology and ethics. Bingley: Emerald Group; 2001. p. 213–38.

Moll J, Eslinger PJ, Oliveira-Souza RD. Frontopolar and anterior temporal cortex activation in a moral judgment task: preliminary functional MRI results in normal subjects. Arq Neuropsiquiatr. 2001;59(3B):657–64.

Moll J, de Oliveira-Souza R, Eslinger PJ, Bramati IE, Mourão-Miranda J, Andreiuolo PA, Pessoa L. The neural correlates of moral sensitivity: a functional magnetic resonance imaging investigation of basic and moral emotions. J Neurosci. 2002;22(7):2730–6.

Moll J, Zahn R, de Oliveira-Souza R, Krueger F, Grafman J. The neural basis of human moral cognition. Nat Rev Neurosci. 2005;6(10):799.

Moll J, Krueger F, Zahn R, Pardini M, de Oliveira-Souza R, Grafman J. Human fronto–mesolimbic networks guide decisions about charitable donation. Proc Natl Acad Sci USA. 2006;103 (42):15623–8.

Moore C, Gino F. Ethically adrift: how others pull our moral compass from true North, and how we can fix it. Res Organ Behav. 2013;33:53–77.

Prehn K, Korczykowski M, Rao H, Fang Z, Detre JA, Robertson DC. Neural correlates of post-conventional moral reasoning: a voxel-based morphometry study. PLoS One. 2015;10(6): e0122914.

Pujol J, Batalla I, Contreras-Rodríguez O, Harrison BJ, Pera V, Hernández-Ribas R, Real E, Bosa L, Soriano-Mas C, Deus J, López-Sola M. Breakdown in the brain network subserving moral judgment in criminal psychopathy. Soc Cogn Affect Neurosci. 2011;7(8):917–23.

Rachul C, Zarzeczny A. The rise of neuroskepticism. Int J Law Psychiatry. 2012;35(2):77–81.

Racine E, Dubljević V, Jox RJ, Baertschi B, Christensen JF, Farisco M, Jotterand F, Kahane G, Müller S. Can neuroscience contribute to practical ethics? A critical review and discussion of the methodological and translational challenges of the neuroscience of ethics. Bioethics. 2017;31 (5):328–37.

Randall DM, Fernandes MF. The social desirability response bias in ethics research. J Bus Ethics. 1991;10(11):805–17.

Rest JR. Research on moral development: implications for training counseling psychologists. Couns Psychol. 1984;12(3):19–29.

Reynolds SJ. A neurocognitive model of the ethical decision-making process: implications for study and practice. J Appl Psychol. 2006a;91(4):737.

Reynolds SJ. Moral awareness and ethical predispositions: investigating the role of individual differences in the recognition of moral issues. J Appl Psychol. 2006b;91(1):233.

Rilling JK, Gutman DA, Zeh TR, Pagnoni G, Berns GS, Kilts CD. A neural basis for social cooperation. Neuron. 2002;35(2):395–405.

Robertson DC. Empiricism in business ethics: suggested research directions. J Bus Ethics. 1993;12 (8):585–99.

Robertson DC, Snarey J, Ousley O, Harenski K, Bowman FD, Gilkey R, Kilts C. The neural processing of moral sensitivity to issues of justice and care. Neuropsychologia. 2007;45 (4):755–66.

Robertson DC, Voegtlin C, Maak T. Business ethics: the promise of neuroscience. J Bus Ethics. 2017;144(4):679–97.

Salvador R, Folger RG. Business ethics and the brain: Rommel Salvador and Robert G Folger. Bus Ethics Q. 2009 Jan;19(1):1–31.

Schwartz MS. Ethical decision-making theory: an integrated approach. J Bus Ethics. 2016;139(4):755–76.

Shenhav A, Greene JD. Integrative moral judgment: dissociating the roles of the amygdala and ventromedial prefrontal cortex. J Neurosci. 2014;34(13):4741–9.

Tenbrunsel AE, Messick DM. Ethical fading: the role of self-deception in unethical behavior. Soc Justice Res. 2004;17(2):223–36.

Thaler RH, Sunstein CR. Nudge: improving decisions about health, wealth, and happiness. New York: Penguin Books; 2009.

Treviño LK. Ethical decision making in organizations: a person-situation interactionist model. Acad Manag Rev. 1986;11(3):601–17.

Treviño LK, Weaver GR. Business ETHICS/BUSINESS ethics: one field or two? Bus Ethics Q. 1994;4:113–28.

Treviño LK, Weaver GR, Reynolds SJ. Behavioral ethics in organizations: a review. J Manag. 2006;32(6):951–90.

Vilares I, Wesley MJ, Ahn WY, Bonnie RJ, Hoffman M, Jones OD, Morse SJ, Yaffe G, Lohrenz T, Montague PR. Predicting the knowledge–recklessness distinction in the human brain. Proc Natl Acad Sci USA. 2017;114(12):3222–7.

Weaver GR, Treviño LK. Normative and empirical business ethics: separation, marriage of convenience, or marriage of necessity? Bus Ethics Q. 1994;4:129–43.

Weisberg DS. Caveat lector: the presentation of neuroscience information in the popular media. Sci Rev Mental Health Pract. 2008;6(1):51–6.

Wolpe PR, Foster KR, Langleben DD. Emerging neurotechnologies for lie-detection: promises and perils. Am J Bioeth. 2005;5(2):39–49.

Young L, Koenigs M. Investigating emotion in moral cognition: a review of evidence from functional neuroimaging and neuropsychology. Br Med Bull. 2007;84(1):69–79.

Young L, Saxe R. The neural basis of belief encoding and integration in moral judgment. NeuroImage. 2008;40(4):1912–20.

Young L, Camprodon JA, Hauser M, Pascual-Leone A, Saxe R. Disruption of the right temporoparietal junction with transcranial magnetic stimulation reduces the role of beliefs in moral judgments. Proc Natl Acad Sci USA. 2010;107(15):6753–8.

Corporate Social Responsibility and Dehumanization

10

Gareth Craze

Contents

Abstract

Corporate social responsibility (CSR) is broadly held up as a crucial component of contemporary business practice. It is defined according to the notion that organizations ought, where possible, to voluntarily recognize and practically mitigate or minimize the socio-environmental impacts of their business activities and that in doing so they will meet the expectations of stakeholders affected by their practices. In spite of this, CSR initiatives are generally attached to the proposition that corporations primarily exist to fulfill their own performance objectives and to satisfy organizational benchmarks, with the additional implication that these aspirations will ultimately take precedence over wider macro-social considerations. The following chapter submits that this conception of CSR reflects the underlying neurological tension between the opposing domains of analytic reasoning and empathic or socioemotional reasoning. Using the opposing domains hypothesis, I propose that current conceptualizations and practices of

G. Craze (✉)
Department of Organizational Behavior, Coaching Research Lab, Brain, Mind and Consciousness Laboratory, Case Western Reserve University, Cleveland, OH, USA
e-mail: gareth.craze@case.edu

J. T. Martineau, E. Racine (eds.), *Organizational Neuroethics*, Advances in Neuroethics, https://doi.org/10.1007/978-3-030-27177-0_10

CSR are antithetical to social and ethical reasoning with respect to neural function and can in turn increase the scope for dehumanization below the threshold of conscious awareness. This in turn calls the ethical dimensions of CSR into question.

10.1 Introduction

Corporate social responsibility (CSR) is widely viewed as a critical feature of contemporary business practice. It is broadly predicated on the notion that organizations should, to the practicable degree possible, voluntarily recognize and seek to attenuate the wider social and environmental impacts of their business activities in order to meet the expectations of those affected by those activities. There are a number of different definitions that scholars and practitioners alike have used to define CSR. These various conceptualizations, sometimes competing, sometimes complementary, have typically done little to clarify exactly how organizational performance in this area might be quantitatively determined and, subsequently, how its various impacts might be benchmarked and assessed (Dahlsrud 2008). Depending on the examination of CSR in question, different social and environmental considerations are used as the category or predictor of interest, such as impacts on the physical environment (DiSegni et al. 2015) and human rights—particularly with respect to already marginalized or beleaguered demographic groups (Claire 2017). Unfortunately, none have proven sufficiently robust or comprehensive to capture the totalized societal outcome that is being aspired to, much less establish which particular CSR strategies or activities are best positioned to drive an organization in the direction of such an outcome.

Despite this prevailing lack of conceptual and practical clarity, CSR is widely viewed as an integral feature of modern organizational life, with the underlying wider implication being that corporations, though primarily driven by economic concerns, create socio-environmental impacts and that where such impacts are negative, voluntary efforts should be undertaken on the part of the organization to offset these impacts. In recognition of the esteem in which social responsibility is held by many of their clients, numerous organizations have elected to focus on one or more broad categories of CSR, including environmental efforts (e.g., the mitigation of the organization's carbon footprint, or other non-negligible environmental impacts, and/or a sense of stewardship and care for the environment); philanthropic efforts (e.g., using the organization's resources to benefit charities or community development initiatives); ethical labor efforts (e.g., fair and equitable treatment of employees and/or an inclination to go above and beyond labor law minimum standards and regulations); and volunteering (e.g., using organizational manpower, absent any expectation of monetary return, to provide human resources to non-organizational efforts) (see Stirling et al. 2016; Wang et al. 2016 for comprehensive reviews of the conceptual history of CSR). These categories, while capturing much of the common threads of CSR across most organizations, are by no means exhaustive, and other organizations have channeled their CSR efforts in other areas still (Lindgreen and Swaen 2010).

Increasingly, the notion that CSR initiatives are a vehicle for organizations to fulfill a wide array of obligations to stakeholders—proximal and distal alike—as well as, where possible, to match or exceed the expectations of the society in which the organization is embedded has become widely subscribed to among relevant scholars and practitioners in the domain of business ethics.

Despite this seeming convergence on the value and necessity of CSR, a range of unresolved issues persist with respect to its theoretical clarity and underlying empirical support. This has subsequently served to further obscure the precise criteria that practitioners ought to adopt with respect to implementing enduring and effective CSR initiatives. The juncture in an organization's history when it elects to adopt a CSR policy very often represents a watershed moment in the evolution of the organization. It represents a recognition on the organization's part (however such a recognition was arrived at) that it must compensate wider society in the act of fulfilling its part in the underlying social contract (Amao 2008). And yet in spite of the significance of adopting such a posture and, ideally, following through with its practical implications in kind, precious little scholarly work has been developed to accurately guide organizational practitioners as to how best to develop CSR initiatives and follow through with their practical implementation (Maignan et al. 2005). Conceptual disagreement abounds, for example, as to whether the shift toward adopting CSR as part of an organization's policy platform should be based on a complete overhaul of the prevailing paradigmatic and operational foundations of the organization or whether a more practical, incremental shift of focus away from pure economic concerns toward those more in line with a broader socio-environmental worldview is the more optimal route to positioning the organization in a more CSR-friendly way (Hart and Milstein 1999). Other disputes around prescription and best practice are, regrettably, similarly common in the literature, and many arguments on both sides of any number of these divides have additionally been yet to be cashed out under empirical examination (Lindgreen and Swaen 2010). This is in spite of the term "corporate social responsibility" now having enjoyed five decades in the limelight.

Furthermore, and at least among scholars, much of the discourse centering on CSR has undergone a movement focusing of its ethical and normative dimensions. Conventionally, most scholars tended to seek to understand CSR at a macro-social level of analysis, zeroing in on the community or other such global entity in which an organization found itself and determining which aspects of a CSR policy affected this specific socio-environmental structure (see Hill et al. 2003). In more recent times, however, this focus has been wheeled back to the organizational level of analysis and has seen a commensurate refocusing on performance outcomes that are central to the organization itself. In keeping with this development, there has been a simultaneous switch in the emphasis scholars have placed on the normative dimensions of CSR—away from an approach largely steeped in wider societal ethical concerns and toward a more utility-driven focus on organization-centric processes and metrics (Lee 2008).

The winds of change that have swept through scholarship in this respect have considerable implications for organizational practice and may further exacerbate the notion, outlined earlier, that organizations have been bedeviled by an overall lack of clarity with respect to how best to put CSR initiatives into practice and, thereafter, assess them as successful or otherwise based on a determinable catalogue of verifiable and relevant criteria. As such, the economic metrics used by the organization to assess for relevant *internal* success may, however unwittingly, become used as a proxy or substitute for determining its success in the realm of CSR. As Du, Bhattacharya, and Sen Du (2010) observe, CSR activities "are driven not just by ideological thinking that corporations can be a powerful and positive force for social change, but more by the multi-faceted business returns that corporations can potentially reap from their CSR endeavors" (p. 8; see also Saeidi et al. 2015). This observation is consistent with Devinney (2009) that "corporations exist to generate economic returns, not to solve societal problems. They live to optimize for themselves (i.e., their near stakeholders: shareholders, managers, employees, suppliers, governments, etc.), not the general public" (p. 49). Complicating matters further is the notion, observed by, e.g., Banerjee (2008), that CSR cannot be determined as being successful or not using purely economic criteria, insomuch as it is ostensibly grounded in an authentic sense of social responsibility. If having a sense of responsible social concern is to be at all meaningful for a corporation, then CSR must transcend its immediate, range-of-the-moment critical and economic concerns.

10.2 The Separation Thesis Through a Neuroscientific Lens

The following book chapter seeks to propose that there is a profound irreconcilability, with respect to the ethical priorities of organizations, between fulfilling organizational performance objectives and achieving beneficial outcomes for wider society. In the extant business ethics literature, this impasse has often been referred to as the separation thesis: a construct drawn in turn from divergent stakeholder theory (Freeman 1999; Harris and Freeman 2008; Sandberg 2008). Central to the separation thesis is the notion that economic and ethical priorities are distinct entities within organizations and in turn require different practical and normative approaches, which have different antecedents in terms of how they ultimately motivate organizations to action.

Harris and Freeman (2008) posit that organizational life is so inherently complex, and so massively steeped in organizational nuance, that the separation thesis is little more than an idealized fantasy, not reconcilable with the everyday realities of corporate life. Nonetheless, in this chapter, I will submit that this separation *can*, in fact, be made sense of provided one adopts a suitably micro-level of analysis. Indeed, the level of analysis I propose to support the central claim of the separation thesis is the neurological level. Specifically, I submit that there is a neurological tension between two distinct networks in the brain, those governing task-based and analytical reasoning on one hand and socioemotional and moral reasoning on the other hand (Rochford et al. 2017), and that through the process of social contagion, such a schism in reasoning can be extrapolated to an organization-wide macro-level of analysis. Moreover, and as I shall outline, purely instrumentalist or economically motivated views of CSR might lead, through differential neural network activity at the collective level, to a propensity to dehumanize people—the very people for whom "socially responsible" corporate policy is intended to be responsible to. This, it might be argued, represents an even more critical separation of ethical and economic concerns than those conceived of by the likes of Sandberg (2008) in championing the separation thesis.

Using a neuroscientific perspective in order to analyze the ethical dimensions of CSR necessitates first using a micro-level of analysis, given that brains are localized to one individual and, as such, contain distinct information about that one individual's own neural and psychological signatures. From there, collective intentionality (Tomasello and Rakoczy 2003) may be used as a vehicle through which a collective totality of individual brain states (i.e., the total cognitive capacity of a collective of individuals deploying their own cognitive capacity in unison) can be used to shape macro-level goals and actions, including those inherent in CSR initiatives. Tomasello and Rakoczy (2003) observe that "Complex social institutions are not individual inventions arising out of humans' extraordinary individual brainpower, but rather they are collective cultural products created by many different individuals and groups of individuals over historical time" and, further, "social and cultural interaction and learning depend fundamentally on the way human individuals understand one another" (pp. 121–122). How individuals ultimately seek to understand one another, and whether they do so using either analytical or socioemotional reasoning, has considerable ramifications for how collective intentionality materializes in practice, as well as the conditions in which social contagion creates a vector whereby collective-level behaviors can be realized as the composite outcome of individual-level brain states.

In the remainder of this chapter, I will sketch an outline of the opposing domains hypothesis: which might be conceptualized as a neuroscientific analogue to the separation thesis, insomuch as it makes a conceptual distinction between task-focused concerns and more humanitarian concerns. The opposing domains hypothesis is grounded in the differential, antagonistic relationship between the brain's task-positive and default mode networks; the various, differing behavioral outputs that may be produced in the wake of activating these distinct networks; and the implications this understanding of brain activity might have for making sense of the ethical and normative dimensions of formulating and implementing CSR policy.

10.3 The Opposing Domains Hypothesis

The opposing domains hypothesis (Jack et al. 2013a, b) is based on the idea that the brain's task-positive network (TPN) and default mode network (DMN) have a neural antagonistic and mutual inhibitory relationship with one another, such that activation of one network usually results in deactivation of the other. The TPN is a network of brain regions (including the insular cortex, dorsolateral and ventrolateral prefrontal regions, insular cortex, and supplementary motor area) which are recruited in response to task-based and attention-demanding stimuli. This network has spatial overlap with both the dorsal attentional system and the frontoparietal control system and is activated when an individual engages in activities that require some combination of deliberate attention and task orientation, such as logical reasoning, mathematical reasoning, general problem-solving, and mechanical reasoning tasks (Jack et al. 2013a).

The DMN comprises a number of the brain's midline regions, including the posterior cingulate cortex, medial prefrontal cortex, and discrete regions of the medial temporal and dorsal medial subsystems (Jack et al. 2013a). There are a range of cognitive functions for which this network has been implicated, including theory of mind, autobiographical memory, self-referential and counterfactual thinking, and, most topically, social evaluation and categorization (Anticevic et al. 2012; Jack et al. 2013a; Uddin et al. 2009). A number of these midline regions are individually recruited when people engage in interpersonal reasoning, such as attributions of personhood and internal mental states to others, though many of these regions also co-engage in concert.

The opposing domains hypothesis holds that the TPN and DMN are anticorrelated—that is, both independent of and suppressive of one another. Jack et al. (2013a) showed, under experimental conditions, that analytical tasks, which recruited the TPN, also deactivated brain regions associated with social reasoning, namely, the DMN, and vice versa such that social reasoning tasks deactivated brain regions associated with analytical reasoning. Moreover, both of these brain networks were shown to be anticorrelated at rest, with this phenomenon also not being explicable in terms of the constraints of the experimental task or other considerations underlying locus of attention, self-referential processing, and so forth.

Individual behavior that results from the activation of either the TPN or DMN reflects a competitive neural relationship between these two networks in terms of their underlying functional connectivity (Kelly et al. 2008). Put another way, the two networks jostle between each other for the utilization of the inputs accrued to the brain through working memory, attention, and the other cognitive systems which undergird the informational substrate that permits all neural network activity to proceed. One is encouraged to read the thoughtful work of, e.g., Mišić and Sporns (2016), for a more comprehensive treatment of the differing mechanisms governing the relationship between individual neural activity and outward behavior, including the most recent work from fields such as neurobiology to underscore this relationship. For the purposes of the present chapter, it should suffice to observe that there are a number of hypothesized, and well-documented, neural pathways which connect brain activity with behavior and that these form the basis of a highly informative literature across the cognitive sciences.

As regards mentalizing—the process used to adopt a phenomenal posture about the subjective mental states of both ourselves and other people—the opposing domains hypothesis can be used to make a distinction between this cognitive faculty and the seemingly related operation of the brain's mirror neuron system (a system for which the broadstrokes of its functionality is reasonably widely known, even among laypeople). The system that the brain deploys in the act of engaging in mentalizing has substantial overlap with the DMN and is recruited in response to stimuli that has a qualitative social dimension to it and particularly stimuli that prompts one to think about the phenomenal experience of other people: their feelings, beliefs, and present emotional state. Conversely, the mirror neuron system—commonly implicated in the act of adopting such a phenomenal posture—is actually more reliably engaged when observing a transient action on the part of another person which has no or limited interpersonal context to it (such as witnessing someone else moving an appendage) (Jack 2013).

10.4 Relationship to the Mind-Body Problem

Within the wider philosophy literature, the so-called "mind-body problem" (the desire, and yet purported inability, to understand conscious mental states in terms of reductionist, naturalistic principles) has been the subject of lively debate and an expansive scholarship (see Westphal 2016, for a comprehensive review of the state of the relevant literature). This conversation has additionally inspired a swathe of focused empirical work in seeking to unpack the most thorny and contentious of both neurobiological and normative issues that seem to hinge on exploring the problem in-depth (see, e.g., Bunge 2014, for an outline of nascent theoretical developments and most informative empirical findings in this space). A detailed history and full overview of the mind-body conversation is beyond the scope of the present chapter. Nonetheless, given that it might have implications for crossing the divide between individual-level neurobiology and collective-level ethics, it bears mentioning that the opposing domains hypothesis might serve as a novel source of insight here.

An irony of the greater interface between philosophy and neuroscience in recent times is that while the mechanisms underlying neural function in general have begun to be more fulsomely understood, this greater understanding has produced increasingly more explanatory gaps of the sort that characterize the mind-body problem in the first place. Indeed, the schism between neurology and phenomenology has arguably become *less* clear than ever when one considers the breathtaking computational complexity that gives rise to individual qualia: how does an array of such seemingly humdrum cellular activity generate the richness of experience inherent in human life? In the present chapter, I will not commit to any of the positions (pure reductionism, property dualism, etc.) that have conventionally characterized the different sides in this space. Invoking the opposing domains hypothesis, fortunately, does not necessitate such a commitment.

Humans have the ability to toggle seamlessly and unconsciously between the TPN and the DMN, something that is representative of healthy general mental function (Hanna 2012). It is a capacity that allows us to transition between domains of reasoning based on relevant cues from the environment that prompt one network or the other into action and one that allows us to make both *analytic* and *empathic* assessments of the nature of phenomenology, for us to understand both that phenomenology itself can be analyzed and that phenomenological states can be empathized with. This is at the heart of what has been termed *the phenomenal stance*; the *physical* capacity of an entity to apprehend the *experiential* states of themselves and others (Jack and Robbins 2012). From this perspective, the supposed explanatory gap presented by the mind-body problem is not an "objective" feature of the world, but rather a reflection of the neural architecture undergirding human subjectivity in two separate domains. Specifically, the TPN is recruited when the brain seeks to understand phenomenal experience in an analytic form, reckoning about the physical brain activity in oneself and others, and the DMN is recruited when the brain engages in interpersonal reasoning, empathically responding to the subjective personal experience of another person through the vehicle of one's own. The process of switching between these two networks is so rapid and seamless that two distinct acts of cognition between two anticorrelated brain networks, objectively speaking, *feel* as though they were one singular instantiation of cognition on a subjective level.

Of course, this perspective on the mind-body problem will not be sufficient to close any important philosophical loops in general. It may, however, in the context of crossing the divide between neurology and normative ethics, shine some helpful light. Namely, it provides an account of the human capacity to engage in instrumental, analytic reasoning about the experiences, personhood, and mental states of other people in spite of also possessing the capacity to empathize with them in a way that is more personally salient and, arguably, more characteristic of the common everyday interpersonal human experience. What may, in many people, outwardly appear to be a lack of empathy toward, or even an unwillingness to recognize the humanity of, another person might actually simply reflect the temporary cognitive occupation of a neural network that is purposed toward instrumentalistic thinking. As we shall see in the following section, this has particular implications for the phenomenon of dehumanization, which is at the crux of the present chapter.

10.5 Dehumanization

Perhaps the greatest significance of the opposing domains hypothesis with respect to assessing the ethical and practical dimensions of CSR, and the relationship between neural activity and subsequent behavior that might characterize these dimensions, is its ability to make sense of the phenomenon of dehumanization. Jack et al. (2013a) demonstrated, under experimental conditions, that the TPN is activated (and the DMN deactivated) when subjects were induced to engage in a process of dehumanizing other people. In doing so, participants were prompted to view people as robotic, mechanistic, animalistic, or otherwise having an absence of the sort of personhood that can be inferred in the context of interpersonal reasoning. The work of Bagozzi et al. (2013) illustrates this phenomenon further, showing that similar neural patterns were shown in individuals exhibiting so-called "Machiavellian" thinking: those who adopt a highly instrumentalistic, opportunistic form of social reasoning devoid of genuine empathy (which would require recruitment of the DMN).

What's more, evidence has suggested that transiently recruiting the TPN in the service of performing ordinary quantitative reasoning tasks can effectively reduce empathic concern and other prosocial behaviors (Bagozzi et al. 2013; Zhong 2011). As a result, when individuals perform what might simply be their ordinary, everyday job roles, they can experience unanticipated deficits in empathy below the threshold of conscious awareness. Given the centrality of an absence of empathy to antisocial cognition in general, and the propensity to dehumanize other people specifically, this should be seen as an especially important consideration in the realm of business ethics. Christoff (2011) and Rochford et al. (2017) have produced thoughtful, comprehensive review articles outlining the complex interplay between the two opposing brain networks and the implications their contextual activation has for the domain of ethical leadership—particularly in light of the potential interference that can occur upon empathic or prosocial behaviors due to engagement in mechanistic or analytic reasoning.

10.6 The Relevance of the Opposing Domains Hypothesis to Corporate Social Responsibility

The emerging field of organizational neuroscience, and related subfields such as organizational neuroethics, has increasingly synthesized many of the findings in social, cognitive, and affective neuroscience with work from fields such as organizational behavior, industrial-organizational psychology, and business ethics (Becker et al. 2011; Robertson et al. 2017). This emerging interdisciplinary approach has sought to establish substantive linkages between our underlying brain biology and the downstream effects that it can produce on human behavior in the organizational realm. While the methodology of neuroscience is based on empirical enquiry, the evidence amassed from neuroimaging studies can produce important normative and ethical questions about human agency and responsibility at multiple levels of analysis, including that of the organization. At the same time, it is important to recognize that while this developing field shows exceptional promise in terms of its ability to inform scholarship in business ethics, it is a nascent area of inquiry, with a number of methodological, theoretical, and ethical objections surrounding its current usage (Farah et al. 2014; Rachul and Zarzeczny 2012). While such disagreements continue to unfold in the scholarly conversation, it is still reasonable to argue that neuroscience perspectives have significant potential for expanding our understanding as to how ethical decision-making operates below the threshold of consciousness and what implications the various subconscious biases which dynamically influence human behavior might have for ethical features of organizational life.

Organizational neuroscience, and the related, nascent field of organizational neuroethics, has encouragingly built on this emerging interdisciplinary approach by generating a range of substantive linkages between human brain biology and its downstream influences on human behavior in the domain of organizational life. Although methodologically speaking neuroscience is based on pure empirical enquiry, much of the evidence it has amassed as a field can contribute crucial normative and ethical framings to questions around normative behavior, policy creation, and best practice in organizations. One must remain vigilant to the reality that while these fields are nascent and aspiring to firm up their own theoretical terrain, they are currently still a source of contention for many scholars, principally with respect to persistent issues around methodology, empirical rigor, and theoretical generalizability.

Nonetheless, neuroscientific perspectives hold significant promise and have great potential for deepening our understanding as to how ethical decision-making occurs at the level of the brain, below the threshold of consciousness, and what ramifications such subsconscious cognitive processes may have for influencing human behavior at the individual level and how this might color and sculpt organizational ethics. In the present context, I submit that using this lens to examine the ethical dimensions of CSR might yield valuable insights into many of the issues surrounding its conception and implementation. Specifically, the opposing domains hypothesis provides a platform upon which the misalignment between firm-level performance objectives and socio-environmental responsibility considerations outlined earlier can be understood in terms of the misalignment between the neural correlates of analytic and socioemotional reasoning and the subsequent scope for dehumanization inherent in this neural opposition.

If any number of the scholars identified earlier (Du Shuili and Bhattacharya 2010; Devinney 2009) are to be believed, and the practical and structural realities of the prevalent, narrowly economic focus endure for corporations, then organizational strategizing will, at its most consequential, be more concerned with utility-based concerns surrounding the increase of shareholder value and maximizing returns on capital. If these claims are true, it appears difficult to reconcile CSR initiatives with anything that might be called "genuine empathy" or a "recognition of personhood" for and of the very people which constitute the "social" entity to which the corporation is ostensibly "responsible" to. If engendering genuine empathy toward people is at odds with profit maximization or any other business considerations that are contingent on achieving certain bottom-line performance benchmarks, then those very people will ultimately be viewed through an instrumental lens. In approaching a CSR initiative through the process of making (or not making) the decision to weigh up social concerns against organizational-economic concerns, the very people comprising the former will be, however unwittingly, reduced to mere "numbers on a balance sheet," whether or not the particular CSR initiative in question is otherwise well-motivated or intentioned toward serving some societal goal. The *humanness* of the people to be affected by the consequence of such a policy is not, at the level of unconscious neural functioning, the consideration when they are comparatively catalogued in a way that establishes their impact on organizational performance.

I propose that such an instrumentalized view of people—one which effectively subjugates their phenomenal qualities as a human being to organizational ends and which is characteristic of CSR as it is currently conceived of and practiced—reveals a TPN-based model of social reasoning, which may consequentially result in dehumanizing behaviors on the part of CSR practitioners. As opposed to viewing the experiences, emotions, mental states, and other aspects of their phenomenal life as a matter of deontological bedrock (which would require activation of the DMN and associated recruitment of one's faculties for empathy), the analytical nature of CSR requires sustained, widespread activation of the TPN, such that CSR practitioners are likely to view those affected by CSR practices as abstractions or quantities. Such sustainment of activity in this network, as outlined earlier, provides fertile ground for the creep of dehumanization, even in contexts where the stimulus in question is a simple performative dimension of one's job. Moreover, the potential for this phenomenon to unfold is a qualitatively different prospect when one considers that the evidence to support it to date has been accrued from small, laboratory study-appropriate sample sizes that are restricted by many of the onerous methodological constraints of neuroimaging scholarship generally. It is an altogether different, and perhaps more sobering, prospect when one extends it to the breadth of an entire corporation or other organizational entity. This is particularly the case when one considers it in light of the transmissive mechanism of social contagion—through which aspects of human cognition and affect can be triggered in one person as a result of the cognitive, affective, and mood-related influence of people around them (Christakis and Fowler 2013). Notably, unethical behaviors have been empirically shown to be amplified in both frequency and effective application as a result of one's embeddedness in a particular social network (Gino et al. 2009). The sum implication here is that TPN activation at the individual level, when occurring in a suitably contagious social nimbus, and even only as a result of seemingly innocuous instances of analytic or mechanistic reasoning in an otherwise social context, could result in a dehumanizing posture being adopted across a collective.

In turn, and in consilience with Devinney (2009), I submit that the raison d'être of corporations—the economic, internal performance-driven considerations which ultimately motivate such organizations, including in the area of CSR—is antithetical to the kinds of social and empathically driven forms of reasoning that are consistent with the activation of the DMN and instead requires a sustained, globalized version of TPN-based reasoning at the collective level. In turn, I submit that to recognize this reality is to recognize the massive increase in scope for dehumanization inherent in CSR as it is currently conceived of and, as a result, calls the ethical dimensions of the entire CSR enterprise into question. Were CSR initiatives to be truly socially responsible, in any meaningful sense of the term, and were they purposed toward an earnest mitigation of any potential societal fallout that could be causally attributed to the organization, then they would not be predicated on an instrumentalist view of the people within their social purview. Any kind of organizational-level weighing-up of the trade-offs between firm-level performance objectives and wider macro-social outcomes is a privileging, at the level of neurological function, of mechanistic problem-solving over socioemotional problem-solving. To reason this way is to adopt an implicit normative stance, lurking beneath conscious awareness, of the relative value of both forms of problem-solving in the domain of social responsibility.

10.7 Implications for CSR Practitioners and Scholars

For those making the most consequential of decisions in organizational settings, there is an ever-looming prospect that many of their actions may result in the harm of others, on a spectrum of harm ranging from basically benign to fatally catastrophic. In such junctures, the difficulty for decision-makers often lies in the moral dilemmas that present themselves and require the full gamut of one's cognitive and affective repertoires to frame these decisions in an ethically salient fashion. In instrumentalizing, and potentially dehumanizing, those who would be burdened with any deleterious outcomes of an enacted CSR policy initiative, such decision-makers may have alleviated themselves of the "burdens" of human proximity and interpersonal empathy that might make such decisions inherently difficult. But this, rather obviously, raises a range of grave ethical concerns. And yet much of CSR is, at a fundamental level, predicated on such seemingly heartless thought processes. It doesn't seem speculative to posit that CSR policy making is not premised on any kind of authentic empathic concern for an affected populace, but rather how such a populace can be practically figured into a form of corporate calculus: numbers, not persons.

It may seem like wishful thinking to hope that CSR practices would ever be sufficiently overhauled so as to move away from this general mind-set and toward a more humanistic conception of best practice. Nonetheless, every effort should be mobilized to thrust the worst excesses of CSR into the scholarly limelight and to subject it to adequately robust conceptual and empirical pressure in order to determine whether it is a valuable feature of corporate life at all, let alone being "socially responsible." Such sentiments, all told, still leave organizations with an open question, and one which has received scarce empirical attention to date. Namely, what exactly can organizations do to offset the most negative effects of their CSR initiatives, particularly given that they are scaffolded on a collectivized preponderance of TPN-dominant, task-oriented, and analytical thinking? While this might seem a tall order to give a satisfying, generalizable answer to, there are nevertheless some promising lines of research which may provide room for hope.

Wang et al. (2014) have demonstrated that many antisocial sentiments and behaviors, of the kind that can be induced through imposing analytic or mechanistically reasoned task demands on participants, can be obviated through the mere exposure of participants to stimuli related to social, community, and family values. This is consistent with work from the business ethics literature, which suggests that organizations which are engaged in high-involvement community interface and engagement practices—those for which the organization engages in sustained consultation and collaboration with stakeholders who are directly affected by their corporate practices—tend to be organizations which have a great potential for positive, enduring impact on their communities (Bowen et al. 2010). Taken together, it does not seem a stretch, in light of the opposing domains hypothesis, to suggest that purposive acts of community engagement across the primary decision-makers in organizations might collectively deactivate individual-level TPN function and in turn reduce the scope for antisocial, dehumanizing outcomes. Essentially, the purposeful, deliberate exposure of ethical decision-makers in organizations (including those setting CSR policy) to the visceral, empathizable personhood of those whom their decisions most acutely impact might go some way to reducing the tendency of such people to entirely subordinate CSR-oriented decisions to purely economic organizational concerns.

Still other research has shown that sustained activation of the DMN is associated with an openness to novelty and possibility and a sense of well-being often centered on a sense of eudaimonia or an aspiration toward human flourishing (Kringelbach and Berridge 2009). This would appear to be in stark contrast to the "hedonic pricing of virtues and vices" (p. 53) taken from the quote that opened this chapter (Devinney 2009), a quote which more accurately epitomizes the prioritizing features of the current organizational landscape with respect to CSR. The major ramification of such research is that to the degree that one can deliberately seek out practices and protocols which engaged their DMN, one can, at the level of the brain, be better positioned to think laterally and openly. In the process, one may benefit from an increased capacity to "think outside the box" and subsequently, as might be hoped, a greater inclination to shift one's focus away from a myopic emphasis on pure organizational performance and economic outcomes and toward a more aspirational, visionary conception of organizational success that takes into account wider socio-environmental growth and prosperity. Mindfulness, meditation training, spending time with kith and kin, being in nature, and spending time thinking in visionary terms are all practices that have been variously shown to engage the DMN (Boyatzis et al. 2014; Josipovic et al. 2012). Leadership training protocols which factor such findings into their developmental apparatus may be able to profitably leverage them in order to effectively repurpose CSR initiatives and other socio-environmentally salient organizational policies while at the same time training their leaders to more effectively hone their interpersonal and empathic faculties at the neural level.

Further still, the phenomenon of social contagion outlined in an earlier section in terms of its inherent dangers for propagating dehumanizing behaviors may also be a vehicle for hope in this area, insomuch as moral behavior (both good and bad) can be transmitted throughout individuals as a function of the collective (MacKenzie et al. 2011). To the extent that any of the DMN-engaging modalities and protocols proposed here are effective and sustainable, they are more likely to be so when they enjoy sufficient subscription and uptake from as many in the organization as possible, particularly those in positions of influence or decision consequence. The pathway of social contagion, which is firmly rooted in the psychological and neural architectures of humankind, provides a route for which such uptake may be realized.

Despite the promise offered by research in this area, and its potential to be extrapolated to the organizational level (Singer 2016), a vigilant skepticism is warranted as to whether this would likely change anything of substance with regard to the prevailing utilitarian conception of CSR. As long as the survival and prosperity of the organization ultimately remains the dealbreaking consideration of CSR, any efforts to embed a more DMN-centered approach to its implementation by organizational decision-makers will likely be doomed to fail. Devinney's (2009) observation that "the socially responsible corporation is a fundamental impossibility" (2009: 53) might prove correct not only as a matter of practical business reality but also of neurological (and subsequently behavioral) reality.

In the meantime, a less fatalistic approach to addressing this proposition should continue to utilize all the available tools of neuroscience in order to shape as informed an understanding as possible of the ethical dimensions of CSR. While foolproof prescriptions remain elusive, a somewhat less fatalistic approach would be to continue to deploy (and refine the deployment of) the full suite of methodological tools available from neuroscience in order to sculpt as robust and informed an understanding as possible of the ethical dimensions of CSR with respect to the function of the brain.

References

Amao O. Corporate social responsibility, social contract, corporate personhood and human rights law: understanding the emerging responsibilities of modern corporations. Aust J Legal Philos. 2008;33:100.

Anticevic A, Cole MW, Murray JD, Corlett PR, Wang XJ, Krystal JH. The role of default network deactivation in cognition and disease. Trends Cogn Sci. 2012;16:584–92.

Bagozzi RP, Verbeke WJ, Dietvorst RC, Belschak FD, van den Berg WE, Rietdijk WJ. Theory of mind and empathic explanations of Machiavellianism: a neuroscience perspective. J Manage. 2013;39(7):1760–98.

Banerjee SB. Corporate social responsibility: the good, the bad and the ugly. Crit Socio. 2008;34:51–79.

Becker WJ, Cropanzano R, Sanfey AG. Organizational neuroscience: taking organizational theory inside the neural black box. J Manage. 2011;31:933–61.

Bowen F, Newenham-Kahindi A, Herremans I. When suits meet roots: the antecedents and consequences of community engagement strategy. J Bus Ethics. 2010;95:297–318.

Boyatzis RE, Rochford K, Jack AI. Antagonistic neural networks underlying differentiated leadership roles. Front Hum Neurosci. 2014;8:114.

Bunge M. The mind–body problem: a psychobiological approach. 1st ed. Amsterdam: Elsevier; 2014.

Christakis NA, Fowler JH. Social contagion theory: examining dynamic social networks and human behavior. Stat Med. 2013;32:556–77.

Christoff K. Dehumanization in organizational settings: some scientific and ethical considerations. Front Hum Neurosci. 2011;144(4):679–97.

Claire MD. Human rights: the emerging norm of corporate social responsibility. In: Human Rights and Corporations. 1st ed. London Routledge; 2017.

Dahlsrud A. How corporate social responsibility is defined: an analysis of 37 definitions. Corp Soc Responsib Environ Manag. 2008;15:1–3.

Devinney TM. Is the socially responsible corporation a myth? The good, the bad, and the ugly of corporate social responsibility. Acad Manag Perspect. 2009;23:44–56.

DiSegni DM, Huly M, Akron S. Corporate social responsibility, environmental leadership and financial performance. Soc Responsib J. 2015;11:131–48.

Farah MJ, Hutchinson JB, Phelps EA, Wagner AD. Functional MRI-based lie detection: scientific and societal challenges. Nat Rev Neurosci. 2014;15:123.

Freeman RE. Divergent stakeholder theory. Acad Manage Rev. 1999;24:233–6.

Gino F, Ayal S, Ariely D. Contagion and differentiation in unethical behavior: the effect of one bad apple on the barrel. Psychol Sci. 2009;20:393–8.

Hanna AJR. The brain's default network and its adaptive role in internal mentation. Neuroscientist. 2012;18:251–70.

Harris JD, Freeman RE. The impossibility of the separation thesis: a response to Joakim Sandberg. Bus Ethics Q. 2008;18:541–8.
Hart SL, Milstein MB. Global sustainability and the creative destruction of industries. MIT Sloan Manag Rev. 1999;41:23.
Hill RP, Stephens D, Smith I. Corporate social responsibility: an examination of individual firm behavior. Bus Soc Rev. 2003;108:339–64.
Jack AI. A scientific case for conceptual dualism: the problem of consciousness and the opposing domains hypothesis. Oxford Stud Exp Philos. 2013;1:1–32.
Jack AI, Robbins P. The phenomenal stance revisited. Rev Philos Psychol. 2012;3:383–403.
Jack AI, Dawson AJ, Begany KL, Leckie RL, Barry KP, Ciccia AH, et al. fMRI reveals reciprocal inhibition between social and physical cognitive domains. Neuroimage. 2013a;66:385–401.
Jack AI, Dawson AJ, Norr ME. Seeing human: distinct and overlapping neural signatures associated with two forms of dehumanization. Neuroimage. 2013b;79:313–28.
Josipovic Z, Dinstein I, Weber J, Heeger DJ. Influence of meditation on anti-correlated networks in the brain. Front Hum Neurosci. 2012;5:183.
Kelly AC, Uddin LQ, Biswal BB, Castellanos FX, Milham MP. Competition between functional brain networks mediates behavioral variability. Neuroimage. 2008;39:527–37.
Kringelbach ML, Berridge KC. Towards a functional neuroanatomy of pleasure and happiness. Trends Cogn Sci. 2009;13:479–87.
Lee MD. A review of the theories of corporate social responsibility: its evolutionary path and the road ahead. Int J Manag Rev. 2008;10:53–73.
Lindgreen A, Swaen V. Corporate social responsibility. Int J Manag Rev. 2010;12:1–7.
MacKenzie C, Garavan TN, Carbery R. Understanding and preventing dysfunctional behavior in organizations: conceptualizing the contribution of human resource development. Hum Resour Dev Rev. 2011;10:346–80.
Maignan I, Ferrell OC, Ferrell L. A stakeholder model for implementing social responsibility in marketing. Eur J Mark. 2005;39:956–77.
Mišić B, Sporns O. From regions to connections and networks: new bridges between brain and behavior. Curr Opin Neurobiol. 2016;40:1–7.
Rachul C, Zarzeczny A. The rise of neuroskepticism. Int J Law Psychiatry. 2012;35:77–81.
Robertson DC, Voegtlin C, Maak T. Business ethics: the promise of neuroscience. J Bus Ethics. 2017;144:679–97.
Rochford KC, Jack AI, Boyatzis RE, French SE. Ethical leadership as a balance between opposing neural networks. J Bus Ethics. 2017;144:755–70.
Saeidi SP, Sofian S, Saeidi P, Saeidi SP, Saaeidi SA. How does corporate social responsibility contribute to firm financial performance? The mediating role of competitive advantage, reputation, and customer satisfaction. J Bus Res. 2015;68:341–50.
Sandberg J. Understanding the separation thesis. Bus Ethics Q. 2008;18:213–32.
Shuili D, Bhattacharya CB. Maximizing business returns to corporate social responsibility (CSR): the role of CSR communication. Int J Manage Rev. 2010;12:8–19.
Singer T. Plasticity of the social brain: from training the mind and heart to a caring society. Harvard Medical School; 2016. http://hdl.handle.net/11858/00-001M-0000-002B-5244-A. Accessed Oct 2016
Stirling JS, Palazzo G, Phillips RA. Historic corporate social responsibility. Acad Manage Rev. 2016;41:700–19.
Tomasello M, Rakoczy H. What makes human cognition unique? From individual to shared to collective intentionality. Mind Lang. 2003;18:121–47.
Uddin LQ, Kelly AC, Biswal BB, Castellanos FX, Milham MP. Functional connectivity of default mode network components: correlation, anticorrelation, and causality. Hum Brain Mapp. 2009;30:625–37.

Wang L, Zhong CB, Murnighan JK. The social and ethical consequences of a calculative mindset. Organ Behav Hum Decis Process. 2014;125:39–49.
Wang H, Tong L, Takeuchi R, George G. Corporate social responsibility: an overview and new research directions. Acad Manage J. 2016;59:534–44.
Westphal J. The mind–body problem. 1st ed. Cambridge: MIT Press; 2016.
Zhong CB. The ethical dangers of deliberative decision making. Adm Sci Q. 2011;56:1–25.

Understanding Unethical Decision-Making in Organizations and Proposals for Its Avoidance: The Contribution of Neuroscience

11

J. Félix Lozano

Contents

Abstract

The aim of this chapter is to propose strategies for improving ethical decision-making in organizations by taking into consideration the contributions of neuroethics research. Our hypothesis is that results from research in neuroscience and cognitive science offer new insight and perspectives on ethical decision-making in organizations.

This chapter has three main parts. The first part is descriptive, and we present the main factors affecting unethical decision-making in organizations. Understanding why good people make or contribute to unethical decisions in organizations is a first step. In the second part, we discuss the contribution of neuroscience research to ethical decision-making. Finally, we propose strategies and actions that managers should take into account to develop a more consistent ethical decision-making model for individuals and organizations.

J. F. Lozano (✉)
Instituto Ingenio (CSIC-UPV), Universitat Politècnica de València, Valencia, Spain
e-mail: jlozan@dpi.upv.es

J. T. Martineau, E. Racine (eds.), *Organizational Neuroethics*, Advances in Neuroethics, https://doi.org/10.1007/978-3-030-27177-0_11

11.1 Introduction

Neuroscience seeks to understand human action in terms of neural mechanisms and physiological interactions that produce cognitive processes and behavior. Understanding how the brain works helps us to understand how human beings think, feel, and act. Spectacular advances in the last two decades have been based in three fields: technological developments that enable us to monitor brain activity (fMRI scanner, CT, EEG, and PET), medical research on patients with brain damage, and the expansion of neuroscience to other fields. One of the more challenging fields derived from neuroscience and cognitive psychology is neuroethics, understood as the neuroscience of moral judgment and behavior (Evers et al. 2017). Knowing more about how our brain and cognitive system works has had a deep impact in the field of moral philosophy on the levels of both theory and practice. On the theoretical level, research in neuroethics is questioning basic ideas like freedom, reason, and responsibility, among others. On the practical level, advances in neuroethics have had a major impact on understanding how people really make moral decisions in everyday life in many fields like business, professions, and organizations (Robertson et al. 2017; Salvador and Folger 2009).

Organizational cognitive neuroscience is the application of neuroscientific methods for analyzing and understanding human behavior in organizations (Butler et al. 2016). The main aim of this field is to apply our understanding of how the brain works, to provide a biologically based (scientific) explanation of human behavior and decisions in organizational contexts. This ambitious objective has been formulated as "(. . .) the study of the processes in the human brain that allow people to understand others, understand themselves, and navigate the social world effectively" (Butler and Senior 2007: 4).

The literature review by Butler and colleagues (Butler et al. 2016; Butler and Senior 2007) identifies three main topics in organizational cognitive neuroscience: (1) neuroscientific studies of economic decision-making, where studies on biological influences (hormone, brain activations, etc.) predominate; (2) neuroscientific studies of decision-making within a marketing context, where hormonal testosterone levels and neuroimaging are the basic scientific data for understanding consumers' decisions; and (3) neuroscientific studies of decision-making within organizations, where a wide range of techniques (hormone monitoring, neuroimaging, EEG, facial morphology, and fluctuating asymmetry) are used to analyze, understand, predict, and manage human behavior in organizations, including leadership. In this chapter, we focus on the last topic, ethical decision-making in organizational contexts, paying special attention to unethical and deviant behavior in organizations.

On ethical decision-making in particular, evidence from neuroethics research covers four broad themes: the neural substrates of normative ethical theories, the role of emotion, the role of intuition, and the specificity of ethical decision-making. Decision-making is a process that integrates cognitive, emotional, and motivational dimensions. Those dimensions are based on neurochemistry, neuroanatomy, and social function, which are key dimensions for explaining relevant moral decisions (Potemkowski 2017).

The objective of this chapter is to identify the main implications of neuroscience research for ethical decision-making in organizations. The first part of the chapter is descriptive and presents the main epistemic, psychological, and contextual challenges from neuroethics research. The second part takes a normative approach and proposes strategies for developing a more consistent ethical decision-making model for individuals and organizations. Finally, we present some practical strategies for moving from words to deeds in ethical decision-making in business organizations. We believe neuroscience can open new perspectives for organizational behavior studies (Becker and Cropanzano 2010; Becker et al. 2011). It may not bring about a revolution, and we should avoid any form of neuro-euphoria (Ashkanasy et al. 2014; Lindebaum and Zundel 2013), but neuroethics certainly represents a significant evolution in the way ethical decision-making is understood in organizations.

11.2 Unethical Decision-Making in Business Organizations: Three Levels of Analysis

One way to study ethical decision-making in organizations is by paying attention to the opposite: unethical decisions and deviant behavior. Everyone in the business world is familiar with the Enron case, the Bhopal tragedy, the Challenger explosion, and the Madoff scandal. Those are just four examples of irresponsible decisions based (mainly) not on bad faith, egoism, or greed but on lack of awareness and sensitivity to the moral dimension of decisions. The struggle to answer the question about why decent people act unethically or cooperate with irresponsible and malevolent organizations is not new. The trigger for this research was the reflection on the atrocities committed by the Nazi system (Arendt 1963; Browing 1993; Milgram 1974), but in recent decades, it has focused also on the corporate world (Bandura et al. 2000, 1996; Moore et al. 2012). The essential questions are why do reasonable, mature, and non-malevolent managers take unethical decisions or cooperate with corrupt organizations? Why do people who value morality commit unethical actions? The answer to these questions can be classified into two main streams: *intentional unethical behavior* when social forces lead individuals to commit unethical actions and *unintentional unethical behavior*, when people are unware of the ethical dimension of their actions (Gino 2015).

Our answer to this question proposes a different classification. We attempt to offer a classification of responses to the question about how decent people make unethical decisions or cooperate with corrupt systems looking at explanations based on three different levels of analysis: epistemic, psychological, and contextual.

(a) The epistemic level of analysis examines the main ideas, concepts, and cognitive assumptions that dominate management theories and discourse. In business and economy, some dominant ideologies (Haase and Raufflet 2017) and mental models (Werhane et al. 2010) frame the thinking and actions of agents in this field. Some years ago, Ghoshal stated that "bad managers' theories are destroying good management practices" (Ghoshal 2005). A year earlier, Galbraith (2004) spoke about "innocent fraud"; since the 1970s, Mintzberg has been speaking out against the "technical" vision of management (Mintzberg 1973, 2004, 2009); and Amartya Sen presented a series criticizing the dominant dogma in economics (Sen 1977). These famous authors focused their criticisms on four basic assumptions of economic and management theories:

First assumption: business is like a physical science, where causality and determinism regulate the relation between individual agents.

Second assumption: knowledge has only an instrumental and control value; its emancipatory and transcendental value is ignored (Habermas 1968).

Third assumption: a positivistic vision of the world where only physical and material phenomena are relevant (only what is measurable and countable is relevant and everything should be computable).

Fourth assumption: reason is a calculation about individual preferences and personal utility. A person is "rational" if he does not reveal any inconsistencies in his choice behavior (Sen 1997).

We can conclude that one source of unethical behavior in organizations is due to epistemic narrow-mindedness with ethical consequences.

(b) The psychological level of unethical decision-making is the second level of analysis for understanding unethical and deviant behavior in organizations. The psychological explanation focuses on four main processes: lack of reflection (Arendt 1963; Milgram 1974; Zimbardo 2007), cognitive bias (Ariely 2012; Kahneman 2011), ethical blindness (Palazzo et al. 2012; Werhane et al. 2010), and self-control depletion (Gino et al. 2011; Mead et al. 2009). The first process—lack of reflection—relates to what German philosopher Hanna Arendt called "*gedankenlosgkeit*" (lack of thinking) and "banality of evil." In her report on Eichmann's trial, she showed that Eichmann decided to reject his capacity for moral reasoning and accepted rules and authority without question (Arendt 1963). She argued that in specific situations, people can be blind to the moral dimension of their actions and act according to the rules or orders of a superior without question. In the same line, research by Stanley Milgram (1974) and Philip Zimbardo (2007) shows how institutions create mechanisms that *force* people to act without moral conscience. One essential aspect at this level is the psychological tendency to moral disengagement (Bandura et al. 2000). How people process information on unethical behavior without feeling bad is a key question for understanding why people make unethical decisions in organizations. According to Moore and colleagues, "The workplace provides ample opportunities for moral disengagement" (Moore et al. 2012: 11).

The second psychological process that influences unethical decision-making is cognitive bias and bounded rationality. Unfortunately, people are less rational than they tend to think and that standard economic theory assumes. As Kahneman (2003, 2011) and Ariely (2008, 2012), among other cognitive psychologists, have showed in recent decades, our cognitive system is not trustworthy. Kahneman and Tversky criticize economic theories for assuming a rational model that is psychologically unrealistic (Kahneman and Twersky

1983; Tversky and Kahneman 1975). They focus on how real people make decisions and conclude that an intuitive system—or cognitive system I (effortless, emotional, fast, automatic, and associative)—dominates many of the decisions we make every day. Cognitive phenomena like framing or heuristic judgments explain the gap between the beliefs people have and the choices they make in daily life. This process is far removed from the idea of rational agent assumed in economic theory and defended by authors like Nobel Laureate G. Becker, who proposed the Simple Model of Rational Crime (SMORC) and stated that dishonest behavior is a result of rational cost-benefit analysis (Becker 1974).

The third process is related to the other two and is called ethical blindness. The problem is not that people make wrong decisions when facing a moral dilemma (bounded rationality) but that people are not aware of moral dilemmas (bounded awareness). To quote Bazerman and Tenbrusel, "Bounded awareness results in the systematic failure to see information that is relevant to our personal lives and professional obligations" (Bazerman and Tenbrunsel 2011: 7). In many cases, especially in large, complex, hierarchical organizations, it is too easy to be blind to the ethical dimension of our actions. We can identify factors that hinder awareness of the ethical dimension at personal and organizational levels. Moral blindness could be a result of personal perceptive limitations (our sensitive capacity is limited and imperfect); of distorted moral self-image or self-deception (everyone tends to think that they are moral, regardless of what they do); and of ideological worldviews (Werhane et al. 2010). At the organizational level, the organizational culture (language, myths, narratives, etc.) and organizational structure (rules, incentives, etc.) are forces that limit the perception of moral wrongdoing (Bazerman and Tenbrunsel 2011; Werhane et al. 2010). The fact is that many people on many occasions act unethically without being aware of it (Gino 2015).

Finally, the last psychological factor is our cognitive capacity for self-control. Relevant research (Gino et al. 2011; Mead et al. 2009) demonstrates that individuals depleted of self-control resources are more likely to behave dishonestly. Gino et al. define self-control as "the psychological capacity that enables people to enact behaviors that are consistent with their long-term goals (e.g., of being an ethical person) and refrain from engaging in behaviors that are driven by short-term, selfish motives" (Gino et al. 2011: 192). They conclude that cognitive overload or physiological states (sleep deprivation, being ill, etc.) reduce our cognitive resources, our capacity for ethical awareness and will power. The consequence is that people with self-control depletion are more likely to engage in unethical behavior.

(c) There is long list of contextual factors that influence decision-making in organizations (Jones 1991; McDevitt et al. 2007; Trevino 1986), but, in this review, we are going to follow the classification presented by Steinmann and Löhr (1994): organizational structure and organizational culture (e.g. soft laws, incentives, roles).

Table 11.1 Explanations for unethical conduct and deviant behavior in organizations

Level of analysis	Sub-process	Key references
Epistemology	Technical vision of management Business like a physical science Knowledge has only instrumental and control value Only the physical and material world is relevant Reason is calculation	Ghoshal (2005) Galbraith (2004) Sen (1997, 1977)
Psychology	Lack of reflection Cognitive bias (bounded rationality) Ethical blindness (bounded awareness) Self-control depletion	Kahneman and Twersky (1983), Kahneman (2011) Ariely (2012, 2008), Gino et al. (2011) Mead et al. (2009, Gino et al. (2011)
Contextual	Structural barriers Cultural barriers	Steinmann and Löhr (1994)

Structural barriers include division of labor, diffusion of decision-making competences, and classical chain of command. Division of labor presents difficulties for ethical decision-making because people lose a broad perspective on the result of their actions and because they see themselves as "a cog in the machine." The separation of decision-making also hinders ethical decisions by creating a hierarchical gap between those who command and those who obey. The classical chain of command and obedience (as in the military) is one of the greatest risks for ethical decision-making. When people are given a very clear limitation of their scope of action and strong pressure to obey a superior, it becomes very difficult to perceive and denounce bad practices.

Steinmann and Löhr (1994) identify some corporate cultural barriers to ethical decision-making. Strong stereotypes or models embody the organization's (sometimes unwritten) rules and values and are a benchmark for people's behavior, especially subordinates. This factor is very closely related to strong group cohesion that puts the organization's interests before those of people, hinders questioning of those interests, and provides immunity against denouncements of bad internal and external practices. Another cultural factor that hinders responsible decision-making is confusion over priorities. When the organization's explicit objectives change frequently and without explanation, or when goals are stated but action takes the opposite direction, it generates much confusion, which prevents people from making ethical decisions. Finally, policies of "isolation" against outside intervention also make ethical decisions difficult for two reasons: first, the external benchmark for actions is lost and, second, it hinders outside scrutiny or oversight that could detect bad practices (Table 11.1).

As we try to argue here, many factors influence human decisions and behavior in an unethical way; some unconscious forces push people toward unethical behavior. Those forces also provide evidence to refute the idea of the person as a rational, self-determined, individual utility maximizer that dominates the traditional paradigm of business decision-making. Neuroethics research is questioning these assumptions and offers a better understanding of unethical decision-making and deviant behavior in organizations.

11.3 Neuroscience Criticism of Dominant Decision-Making Models in Business Organizations

Advances in neuroscience and cognitive psychology offer a deeper and more rigorous understanding of our mind and behavior, thereby helping us to understand our very identity (Levy 2008). Noninvasive methods for monitoring brain activity (computed tomography, nuclear magnetic resonance, electroencephalography, positron emission tomography, etc.) offer relevant insights for understanding how people think, feel, and act. Especially relevant for ethical decision-making are neurobiochemistry data. The roles of dopamine and serotonin in risky decision-making; the roles of noradrenalin and serotonin in activating or suppressing emotions; and the role of dopamine level in concentration and attention are well established (Potemkowski 2017; Robertson et al. 2017; Salvador and Folger 2009). The neuroanatomy of decision-making confirms that the orbitofrontal cortex is responsible for controlling the spontaneous response to positive stimuli and acting based on a long-term perspective. The dorsolateral prefrontal cortex is responsible for processing, integrating, evaluating, and categorizing information and planning, controlling, and responding to stimuli. Finally, the anterior cingulate cortex is essential for deciding in uncertainty situations, when we have to select between two attractive options (Gold and Shadlen 2007; Hallsson et al. 2018; Potemkowski 2017).

These conclusions present new insights into the theory of decision-making in business and organizations. Following the structure of the previous section, we attempt to present the specific changes for epistemology and psychology. This classification seeks to clarify the specific contribution of neuroscience to our understanding of decision-making theory while taking into account the difficulty of attempting to separate the epistemological and the psychological dimensions.

11.3.1 Epistemology

The idea of reason used by dominant economic theories has been superseded by recent neuroscientific research. The basic idea of economic rationality based on the utility maximization function as a framework for modelling economic decisions is called into question on two essential points: reason is not only calculus nor is it individual. According to Nida-Rümelin, “Thus, the theory of practical rationality can be constructed as a theory of good reasons for action" (Nida-Rümelin 1997: 4). We can identify three types of "good reasons" for action: the first is consequentialist reason, when we make decisions to achieve a desirable objective; promise keeping is another type of reason for acting, without paying attention to the specific consequences of the action; and, finally, acting according to principles is a good reason for acting. The idea of good reasons for acting inevitability involves intersubjectivity, thereby negating the individualistic concept of reason. Making moral judgments "(. . .) demands that good and evil always be evaluated by reason, thus through concepts which allow of universal communication and not through mere sensations (which are limited to individual subjects and their susceptibility) (. . .)" (Kant 1787, Ak. v, 58). This idea of universal communication was also presented in the *Beantwortung der Frage: Was ist Aufklärung?* (Kant 1784) and was the seed of the publicity principle (García-Marzá 2012) developed two centuries later by discursive ethics. According to Habermas (1981) and Apel (1973), giving reasons—dialogue and argumentation—is a specific human capacity that involves, among others, recognizing others, listening to their arguments, following rules of language, compromise in telling the truth, and publically presenting one's own ideas. Intersubjectivity, in this sense, is a specific human capacity that enables communication, cooperation, and ethical reflection. Our reason is, by nature and definition, dialogical.

Moreover, neuroscience research demonstrates that reason is not pure and that intuitions and emotions play a relevant role in decision-making and especially in ethical decision-making. Intuitions find their origin in primary emotions (generated in the limbic system circuitry) and secondary emotions (generated in the prefrontal cortex) (Damasio 1994). Emotions, according to evolutionary psychology, enable adaptive responses to the environment and are the result of thousands of years of evolution (Greene 2013). The classical article by Haidt "The Emotional Dog and Its Rational Tail: A Social Intuitionist Approach to Moral Judgment" (Haidt 2001) was a revolution in the theory of moral judgment. Although some of the conclusions are controversial, it was a strong defense of intuition in moral judgment. He states that in moral questions, people have intuitions that generate moral judgments and that reason comes after this judgement, as a post hoc explanation or justification of this previous, emotive, and unreflective moral judgment. The main conclusion of this intuitionist perspective of moral judgement is that our reason is not so independent and powerful as we used to think; it is strongly influenced by intuitions and emotions generated through evolution as a biological response to the environment. Intuitionism is controversial and has been strongly criticized (Cortina 2011; Dubljević and Racine 2014; Dubljević 2017), but it has the positive effect of focusing attention on the role of intuition and emotions in moral decisions. The negative effect has been to generate an interpretation of two types of moral judgement, intuitive (deontological) and reflexive (utilitarian), which is superficial and philosophically not accurate (Crockett 2013, 2016; Everett et al. 2016).

Human beings as self-governing agents is the second epistemic topic that we would like to broach in the light of neuroscience research. One of the central assumptions of economic decision-making theory is that a person is a rational agent that has some preferences and decides what to do according to those preferences. The idea is that "I decide what to do," where "I" is an agent with clear preferences and capacities to achieve them; he or she has free will and self-governing capacity. According to Galnnon, "The will is free when one has the capacity to recognize and react appropriately to reasons in forming and executing intentions in actions that accord with one's considered desires and values" (Galnnon 2015: 5). Neuroscience research demonstrates that our capacity for acting coherently with our objectives is not as strong as we tend to think. Our will and self-control are weaker than we would admit. Therefore, our capacity to postpone reward, or our ability to resist doing things we think we should not do or not do what we think we should do, is very limited (Gino et al. 2011; Mead et al. 2009; Tenbrunsel and Messick 2004). Neuroscience calls into question this ideal of free will and self-governing agent from two perspectives: one is philosophical and the other psychological.

From the philosophical perspective, the questioning of free will is often based on Benjamin Libet's experiment paradigm (Libet 1999). He demonstrated that before the conscious decision to move the hand, our brain has already decided to do that; and therefore, the volitional process is initiated unconsciously. "Libet-style experiments indicate that the awareness of a decision to act is preceded by brain activity, such as a readiness potential" (Mecacci and Haselager 2015: 328). Such interpretations have expanded and reinforced the idea that self-determination is an illusion. However, Libet's paradigm is highly controversial and presents some weaknesses. After reviewing 48 Libet paradigm articles, Saigle, Dubljević, and Racine conclude that few experiments have been able to produce similar results, they do not establish the relation between neural activity and awareness of action, and essentially they make a basic conceptual mistake when identifying voluntary actions with finger flexions (Saigle et al. 2018).

From the psychological perspective, the influence of factors such as reward, social pressure, anchors, and bias that "prevent" us from doing what we judge to be the best is well documented (Ariely 2008; Caruso and Gino 2011; Gino et al. 2011; Mead et al. 2009). Weakness of will or lack of self-control has been called, since Aristotle, *akrasia* (Kalis et al. 2008). From the scientific perspective, it is astonishing why people do not act according to their values and ideas in vital questions regarding health or money. One relevant example of self-governing weakness is corruption. According to Ariely (2012) and Werhane et al. (2010), many people participate in corrupt practices without noticing it, in fact, on many occasions they believe they are acting fairly. These authors demonstrate how interest, incentives, favors, and rewards affect our perception of reality and the evaluation of our own actions. In conclusion, we are much more influenced and less self-determining agents than we think.

The third relevant epistemic concept that neuroscience has challenged is the objectivity of the external world. A positivistic approach predominates in main-stream theories of business and management that also influences ethical decision-making. The dominant "reducible vision of business to a kind of physics" (Ghoshal 2005: 77) has relevant consequences for the praxis: attention is only paid to data that can be operationalized, measured, and counted. This assumption is inappropriate for two reasons: our bounded perception and our transcendental aspirations.

Neuroscience research has shown that our perception capacity is limited and unreliable. As Gold and Shadlen (2007) point in the context of sensory-motor tasks, perception is the first step in the decision-making process, and this perception includes sensory evidence and the evaluation of stimuli; but the problem is that perception and sensory evidence are sometimes perceptual illusions. Kahneman explains this unreliable perception capacity by the idea of "accessibility" and how our perception is determined by expectation, not by pure objective presence (Kahneman 2003). Moreover, perceptual illusions generate cognitive illusions (Kahneman 2011). Ariely calls this lack of objectivity "relativity." He says that our perception and experience are always and inevitably evaluated in comparison to the context and that we make decisions based on comparisons. "We have to accept that relativity is everywhere, and that we view everything through its lens—rose-colored or otherwise" (Ariely 2008: 8). In essence, the framing effect, confirmation bias, narratives, and mental models not only affect how we think but also how we perceive the world. "Perceiving is as much about acting on the environment as it is about receiving signals from it" (Damasio 1994: 25).

The second argument against the idea that only the material world is relevant is the transcendental aspiration of human beings. History, anthropology, and evolutionary psychology concord that believing in someone or something "metaphysic" (beyond physics) is a specific and essential trait of the human species (Aristotle 1986). According to Haidt, this transcendental aspiration is based on the capacity of the human mind to perceive what he called "divinity" (Haidt 2001), and it is probably an evolutionary adaptation for binding groups together. Thus not only the material and objective world matter (Lancaster 2016; Walach et al. 2011).

We can sum up the previous paragraph stating that recent neuroscience research has called into question three traditional mainstream essential concepts that dominate the decision-making paradigm in business and organizations: reason, self-governing agent, and objective world.

11.3.2 Psychology

Neuroscience is helping us to understand how we act, feel, and think and, in essence, how we understand our selves. Traditional perspectives in behavioral economics and economic psychology have been seriously questioned by results of neuroscience research. Looking back at the key question in our chapter (why decent people act unethically and cooperate with irresponsible organizations), we would like to focus on three essential contributions from neuroscience. Those contributions question the predominant idea of human psychology in business and economic decision-making.

The first topic concerns opposing conceptions of human nature: individualism versus cooperation. According to the mainstream view in business, decision-making is a process for achieving the desired output for the individual actor. It is the economic actor, who, after evaluating different options, chooses the preferred option according to her/his objective. It is an individual calculation process. Cooperation is only a second-best option for achieving one's own desires. Nevertheless, the key point here is not that it is a descriptive theory; it is a normative one. From the standpoint of the liberal economy, the ideal person is a *homo economicus*, who in a natural context fights for his or her own individual interest. It is the essence of human nature to compete for one's own best interest in a scarce resource situation, and that is what we *ought* to do.

According to evolutionary psychology, behavioral genetics, anthropology, and cognitive neuroscience, this perspective is wrong. It is not new to recognize the human species as a social animal (*zoon politikon* according to Aristotle), but what is new is to have scientific and biological evidence of this socialization capacity. "Human beings are heavily influenced by their social setting *because of their biology*[1]" (Becker and Cropanzano 2010: 939). According to Greene, "Biologically speaking, humans were designed for cooperation, but only with some people[2]" (Greene 2013: 42). That means that family ties and group membership are the natural context for cooperation. Emotions like concern for others (empathy), direct reciprocity (anger, disgust, and gratitude), commitment to threats and promises (vengeance, shame, loyalty), reputation (embarrassment), and indirect reciprocity (righteous indignation) are basic for group cooperation, and they have a biological function. In conclusion, we are by nature a cooperative species, not an individualistic one (Tomasello 2009). Rather than *homo economicus*, we are, ontogenetically and phylogenetically, *homo reciprocans* (Cortina 2012; Conill 2006).

The second topic is the dialectic between egoism and altruism. Following the argument presented in the previous paragraph, the dominant perspective in business and economic organizations is that the economic agent is egoist. Thus, people always act according to the best possible option for achieving what they want. However, as Nobel Laureate economist Sen wrote more than four decades ago:

> Economic theory in this, as well as in some other fields, tends to suggest that people are honest only to the extent that they have economic incentives for being so. This is a homo economicus assumption which is far from being obviously true, and which needs confrontation with observed realities. (Sen 1977: 332)

Neuroscience research supports Sen's thesis. Especially relevant are the experiments based on the ultimatum game (Sanfey et al. 2003). When people receive an offer, their decision to accept it is not only based on cold utilitarian calculus. If people behaved like purely self-interested agents, they would be ready to accept any offer that improved their current situation. The attitude of looking beyond personal gain applies not just to economics but also to other valuable topics. Two interesting empirical studies by Crockett et al. (2015) and Story et al. (2015) reach the same conclusion: "Surprisingly, most people sacrifice more money to reduce a stranger's

[1]Highlighted in the original

[2]Highlighted in the original

pain than their own pain" (Crockett et al. 2015: 17320). Another example of the altruistic trait of human psychology concerns not the willingness to help other people but to punish them. "Altruistic punishment" consists in the fact that we are willing to pay a personal cost to punish people who put their self-interest before the group interest.

The last topic is the conflict between emotion and reason in business organizations. Many decades ago, Damasio (1994) unveiled Descarte's error and presented arguments and scientific evidence that reason and emotions are complementary parts of the cognitive system that allows us to live. He showed that, biologically speaking, primary and secondary emotions are the basis for reasoning and that there is no reasoning without emotion and feeling. Using the same empirical experiments presented in the previous paragraph, people in the ultimatum game accept or reject offers based on emotions like feelings of humiliation, indignation, or revenge (Sanfey 2009). One important aspect of this experiment is that when people receive what they consider an unfair offer, they accept it, if it was made by a computer program, but reject it if it was made by another person.

Another clear example of the importance of emotions in decision-making and with relevant implications for organizations is the answer to personal and impersonal dilemmas. The trolley dilemma is a good example for acknowledging the limitation of a rational decision without emotions. Most people are willing to turn the wheel to kill one person instead of five, but the same people are not prepared to push a fat man down the bridge to the tracks (Edmons 2014). We can present many explanations for this observation, but the first one is that we are more emotionally implicated when pushing someone directly than when turning a wheel or mechanism.

We can conclude that the dichotomy between reason and emotion in decision-making and moral judgement is strongly debated. In fact, some authors (Damasio 1994; Woodward 2016) state that the classical distinction between reason and emotion is wrong, from both a philosophical perspective and a physiological one. "(...) the moral requirements that we find appealing and that are suitable for regulating our lives will reflect facts about our affective and motivational commitments, as well as our capacities for reasoning" (Woodward 2016: 113).

The general conclusion of these studies is that human beings are by nature cooperative, altruistic, and emotional and that the dogma of *homo economicus* that dominates the business world is neither real, nor right, nor fair.

11.4 Conclusions: Strategies for Developing Ethical Organizations

Our general conclusion is in ways that neuroethics research helps us to understand why people behave unethically and challenges some basic philosophical and psychological assumptions of business and professional practice. Neuroethics questions the basic assumptions that people's behavior is individualistic and egoistic; it also criticizes the traditional rationalistic perspective where rationality is understood essentially as a strategic calculation for achieving subjective preferences; and finally, it also questions the traditional radical division between rational and emotional, where emotional is an uncontrollable dimension of human behavior.

The neuroscience contribution to understanding ethical or unethical decision-making in business goes far beyond mere criticism of mainstream rationalistic models (Jones 1991; Trevino 1986); it generates some concrete proposals for ethical decision-making that integrate neuroscience contributions. We can identify at least four relevant proposals for ethical decision-making models that incorporate results from neuroscience research.

One of first models was the "sensemaking intuition model" presented by Sonenshein (Sonenshein 2007). The basic assumption in Sonenshein's model is that individuals construct moral issues intuitively from social stimuli in equivocal and uncertain environments, and this construction is affected by different factors like expectation and motivational drivers (personal) and social anchors and representations of mental models (collective). Developing moral intuitions about moral issues in organizations and fostering plurality in discussions of ethical issues in organizational contexts are some interesting practical recommendations of this model that we take into account in our proposal.

Woiceshyn (2011) proposes another relevant model for ethical decision-making in business in which reasoning and intuition interact through forming, recalling, and applying moral principles necessary for long-term success in business. This model is a reformulation of Sonenshein's (2007) model, but here the key concept is "integration by essential," which means that in the decision-making process, the relationship between the conscious and subconscious/intuitive level of information processing is dialectic and integrated by the essential traits of the observed phenomenon.

The third model we would like to comment on here is the proposal from Thiel et al. (2012). They consider sensemaking to be the key process for ethical decisions, and this process is influenced by personal, situational, and environmental constraints. What leaders can do to build sensemaking is to develop four strategies: emotion regulation, self-reflection, forecasting, and information integration.

Similarly, Schwartz (2016) proposes a reformulation of an ethical decision-making model that integrates two broad approaches: the rationalist and intuitionist/sentimentalist. His comprehensive model has two major components: the process and the factors.

Taking into consideration our previous analysis, we would like to propose strategies and actions that managers should take into account for preventing deviant behavior in organizations and for developing a more consistent ethical decision-making model for the personal and organizational realms. All the following proposals share three basic assumptions: the need to change basic ideas and "dogmas" that dominate business theory and practice; the need to integrate emotional and intuitive dimensions in decision-making processes; and the urgent need to develop organizational ethical infrastructures (ethical codes, ethical committees, ethical auditing, etc.) that influence personal behavior.

If we want to increase the ethical level of our organizations and ensure that people working in them behave in a more responsible way and make better decisions, we have to focus our efforts at two levels: personal and organizational. Following Kant, we can say that developing an ethical infrastructure in an organization without considering personal will is to be blind; but focusing only on personal character while ignoring the organizational context is naïve.

At the personal level, we suggest three broad strategies for developing ethical organizations:

(a) Questioning dominant narratives and fostering self-critical perspectives on dogmas and mental models that govern business organizations. As Ghoshal and other authors (Haase and Raufflet 2017; Lozano and Escrich 2017; Martineau et al. 2017) point out, ideologies play a vital role in business and organizations. We have to be aware of that and critically consider their foundations and implications. One way to achieve this could be by participating in open critical dialogue with people with different backgrounds and experience. This critical dialogue can overcome the confirmation bias tendency and foster moral imagination. This culture of questioning and criticism should start in management education (Lozano 2012).
(b) Developing moral intuitions about moral issues in organizations. One of the essential traits of managerial work is speed; and speed is incompatible with reflection. As neuroethics shows, many of our moral decisions are intuitive and quasi-automatic. These intuitive responses to moral questions are based on emotional reactions developed during years of personal maturation and centuries of cultural evolution. It could be a good strategy to develop moral intuitions—or habits to quote Aristotle—in organizations through repetitive clear responses to dilemmas and through emotional linkage between (justified) norms and actions.
(c) Revaluation of the emotional dimension. Although we defend the ADC framework of moral judgment proposed by Dubljević and Racine (Dubljević and Racine 2014) and share the need to overcome the interpretation of the dual-process model in neuroethics (Dubljević 2017; Hallsson et al. 2018), special attention must be paid to the emotional dimension of ethical decision-making in organizations. It is important for managers to be aware of three clear statements: emotions are always present in their decisions; emotions can be positive; and emotions can and should be evaluated and controlled. One concrete recommendation for that could be try to personalize problems and decisions. Our technological world, and especially the organizational context, tends to be impersonal and neutral. However, the fact is that every decision has human implications. We have to avoid the abstract and impersonal dilemmas presented in many models and transform them into personal dilemmas with people suffering or enjoying some consequences.

At organizational level, we would like to propose two strategies:

(a) Develop ethical reward systems. One of the best understood languages in organizations is the economic reward system. If you are paid for doing something, that means it is important. There is often no coherence between what the organizational discourse affirms, the ethical code, and the reward system. Moreover, the reward system is an expression of organizational justice.

(b) Develop an ethical infrastructure. The impact of ethical codes on organizational reputation and culture is well known (Adams et al. 2001; Erwin 2011; Kaptein and Schwartz 2008). But, an ethical code is not enough. An ethical code is like a skeleton; it needs fleshing out to become a living body. Ethical committees, ethical hotlines, and corporate social responsibility reporting and training programs are essential parts of an ethical organization and contribute to institutionalize an ethical culture. An ethical infrastructure helps people make ethical decisions in three ways: it clarifies what is acceptable and what is not, establishes processes for dealing with ethical dilemmas, and makes people think about ethics (Fernández and Camacho 2016; Tenbrunsel et al. 2003).

We are human beings acting in specific contexts. Our decisions and behavior are influenced by ideas, biological factors, cultural rules, and environment, among others. Neuroscience helps us to understand all of this. There is still a long way to go, and in the coming years, cognitive neuroscience research will likely offer incredible results. We will be responsible for evaluating them and integrating them into moral philosophy and organizational theory and practice. But, we also have a duty to take advantage of cognitive neuroscience to change decision-making in organizations today.

Acknowledgements This research is part of the Project FFI2016-76753-C2-1-P, financed by the Spanish Ministry for Science, Innovation and Universities; and is part of the Excelence Research Group PROMETEO/2018/121 by Generalitat Valenciana.

References

Adams JS, Tashchian A, Shore TH. Codes of ethics as signals for ethical behavior. J Bus Ethics. 2001;29(3):199–211.

Apel KO. Transformation der Philosophie. Frankfurt am Main: Surhkamp; 1973.

Arendt H. Eichman in Jerusalem: a report on the banality of evil. New York: Viking; 1963.

Ariely D. Predictably irrational: the hidden forces that shape our decisions. New York: Harper-Collins; 2008.

Ariely D. The honest truth about dishonesty: how we lie to everyone-specially ourselves. New York: Harper-Collins; 2012.

Aristotle. The complete works of Aristotle: The revised Oxford translation. Baumrin, Bernard H. Metaphilosophy. 1986;17:350–1.

Ashkanasy NM, Becker WJ, Waldman DA. Neuroscience and organizational behavior: avoiding both neuro-euphoria and neuro-phobia. J Organ Behav. 2014;35:909–19.

Bandura A, Barbaranelli C, Caprara GV, Pastorelli C. Mechanisms of moral disengagement in the exercise of moral agency. J Pers Soc Psychol. 1996;71(2):364–74.

Bandura A, Caprara GV, Zsolnai L. Corporate transgressions through moral disengagement. J Hum Values. 2000;6(1):57–64.

Bazerman MH, Tenbrunsel AE. Blind spot. Why we fail to do what's right and what to do about it. Princeton: Princeton University Press; 2011.

Becker GS. Crime and punishment: an economic approach. In: Becker G, Landes W, editors. Essays in the economics of crime and punishment. New York: Columbia University Press; 1974.

Becker WJ, Cropanzano R. Organizational neuroscience: the promise and prospects of an emerging discipline. J Organ Behav. 2010;31:1055–9.

Becker WJ, Cropanzano R, Sanfey AG. Organizational neuroscience: taking organizational theory inside the neural black box. J Manage. 2011;37(4):933–61.

Browing CR. Ordinary men. Reserve police battalion 101 and the final solution in Poland. New York: Harper Perennial; 1993.

Butler MJR, Senior C. Toward an organizational cognitive neuroscience. Ann NY Acad Sci. 2007;1118(0):1–17.

Butler MJR, O'Broin HLR, Lee N, Senior C. How organizational cognitive neuroscience can deepen understanding of managerial decision-making: a review of the recent literature and future directions. Int J Manag Rev. 2016;18(4):542–59.

Caruso EM, Gino F. Blind ethics: closing one's eyes polarizes moral judgments and discourages dishonest behavior. Cognition [Internet]. Elsevier; 2011 Feb 1 (cited 2018 May 13);118 (2):280–5. Available from https://www.sciencedirect.com/science/article/pii/S0010027710002672?via%3Dihub

Conill J. Horizontes de economía ética. Madrid: Tecnos; 2006.

Cortina A. Neuroética y Neuropolítica. Sugerencias para la educación moral. Madrid: Tecnos; 2011.

Cortina A. Guía Comares de Neurofilosofía Práctica. Granada: Comares; 2012.

Crockett MJ. Models of morality. Trends Cogn Sci [Internet]. 2013;17(8):363–6.

Crockett MJ. How formal models can illuminate mechanisms of moral judgment and decision making. Curr Dir Psychol Sci. 2016;25(2):85–90.

Crockett MJ, et al. Harm to others outweighs harm to self in moral decision making. Proc Natl Acad Sci. 2015;112(4):E381. https://doi.org/10.1073/pnas.1424572112.

Damasio A. Descartes' error. Emotion, reason, and the human brain. New York: Penguin Books; 1994.

Dubljević V. Is it time to abandon the strong interpretation of the dual-process model in neuroethics? In: Racine E, Aspler J, editors. Debates about neuroethics. Advanced in neuroethics. New York: Springer; 2017. p. 129–40.

Dubljević V, Racine E. The ADC of moral judgment: opening the black box of moral intuitions with heuristics about agents, deeds, and consequences. AJOB Neurosci. 2014;5(4):3–20.

Edmons D. Would you kill the fat man? The trolley problem and what your answer tell about right and wrong. Princeton: Princeton University Press; 2014.

Erwin PM. Corporate codes of conduct: the effects of code content and quality on ethical performance. J Bus Ethics. 2011;99(4):535–48.

Everett JAC, Pizarro D, Crockett MJ. Inference of trustworthiness from intuitive moral judgments. J Exp Psychol Gen Everett 2016;145(6):772–87. Available from file:///C:/Users/mbesw/Downloads/SSRN-id2726330.pdf

Evers K, Sales A, Farisco M. Theoretical framing of neuroethics: the need for a conceptual approach. In: Racine E, Aspler J, editors. Debates about neuroethics. New York: Springer International; 2017. p. 89–107. https://doi.org/10.1007/978-3-319-54651-3.

Fernández JL, Camacho J. Effective elements to establish an ethical infrastructure: an exploratory study of SMEs in the Madrid region. J Bus Ethics. 2016;138(1):113–31.

Galbraith JK. The economics of innocent fraud. Truth for our time. New York: Penguin Books; 2004.

Galnnon W. Free will and the brain. Neuroscientific, philosophical and legal perspectives. Cambridge: Cambridge University Press; 2015.

García-Marzá D. Kant's principle of publicity: the intrinsic relationship between the two formulations. Kant-Studien. Philos. Zeitschrift der Kant-Gesellschaft. 2012;103(1):96–113.

Ghoshal S. Bad management theories are destroying good management practices. Acad Manag Learn Educ. 2005;4(1):75–91.
Gino F. Understanding ordinary unethical behavior: why people who value morality act immorally. Curr Opin Behav Sci. 2015;3:107–11.
Gino F, Schweitzer M, Mead N, Ariely D. Unable to resist temptation: how self-control depletion promotes unethical behavior. Organ Behav Hum Decis Process. 2011;115(2):191–203.
Gold JI, Shadlen MN. The neural basis of decision making. Annu Rev Neurosci. 2007;30(1):535–74. https://doi.org/10.1146/annurev.neuro.29.051605.113038.
Greene J. Moral tribes. Emotion, reason and the gap between us and them. New York: Penguin Books; 2013.
Haase M, Raufflet E. Ideologies in markets, organizations, and business ethics: drafting a map: introduction to the special issue. J Bus Ethics. 2017;142(4):629–39.
Habermas J. Erkenntnis Und Interesse. Hamburg: Felix Meiner Verlag; 1968.
Habermas J. Theorie des kommunikativen Handelns. Bd. 1: Handlungsrationalität und gesellschaftliche Rationalisierung; Bd. 2: Zur Kritik der funktionalistischen Vernunft. Frankfurt am Main: Surhkamp; 1981.
Haidt J. The emotional dog and its rational tail: a social intuitionist approach to moral judgement. Psychol Rev. 2001;108(4):814834.
Hallsson BG, Siebner HR, Hulme OJ. Fairness, fast and slow: a review of dual process models of fairness. Neurosci Biobehav Rev. 2018;89:49–60. https://doi.org/10.1016/j.neubiorev.2018.02.016.
Jones TM. Ethical decision making by individuals in organizations: an issue-contingent model. Acad Manag Rev. 1991;16(2):366–95.
Kahneman D. Maps of bounded rationality: economics psychology for behavioral. Am Econ Rev. 2003;93(5):1449–75.
Kahneman D. Thinking fast and slow. New York: Farrar, Straus and Giroux; 2011.
Kahneman D, Twersky A. Choices, values, and frames: APA Award Address; 1983.
Kalis A, Mojzsch A, Schweizer TS, Kaiser S. Weakness of will, akrasia, and the neuropsychiatry of decision making: an interdisciplinary perspective. Cogn Affect Behav Neurosci. 2008;8(4):402–17.
Kant I. Beantwortung der Frage: was ist Aufklärung. (1994). Göttingen: Vandenhoeck and Ruprecht; 1784.
Kant I. Kritik der praktischen Vernunft. Frankfurt am main: Surhkamp; 1787.
Kaptein M, Schwartz MS. The effectiveness of business codes: a critical examination of existing studies and the development of an integrated research model. J Bus Ethics. 2008;77(2):111–27.
Lancaster BL. Spirituality and cognitive neuroscience: a partnership for refining maps of the mind. Bern: Springer International; 2016. p. 95–106.
Levy N. Introducing neuroethics. Neuroethics. 2008;1:1–8.
Libet B. Do we have free will? J Conscious Stud. 1999;6(8):47–57. Available from http://www.ingentaconnect.com/content/imp/jcs/1999/00000006/F0020008/966
Lindebaum D, Zundel M. Not quite a revolution: scrutinizing organizational neuroscience in leadership studies. Hum Relat. 2013;66(6):857–77.
Lozano JF. Educating responsible managers. The role of university ethos. J Acad Ethics. 2012;10:213–26.
Lozano JF, Escrich T. Cultural diversity in business: a critical reflection on the ideology of tolerance. J Bus Ethics. 2017;142(4):679–96.
Martineau JT, Johnson KJ, Pauchant TC. The pluralist theory of ethics programs orientations and ideologies: an empirical study anchored in requisite variety. J Bus Ethics. 2017;142(4):791–815.
McDevitt R, Giapponi C, Tromley C. A model of ethical decision making: the integration of process and content. J Bus Ethics. 2007;73(2):219–29.
Mead NL, Baumeister RF, Gino F. Too tired to tell the truth self control resource depletion and dishonesty. J Exp Soc Psychol. 2009;45(3):594–7.

Mecacci G, Haselager P. A reason to be free: operationalizing free action. Neuroethics. 2015;8:327–34.

Milgram S. Obedience to authority: an experimental view. New York: Harper and Row; 1974.

Mintzberg H. The nature of managerial work. New York: Harper & Row; 1973.

Mintzberg H. Managers, not MBA. San Francisco: Berret-Koehler; 2004.

Mintzberg H. Managing. San Francisco: Berret-Koehler; 2009.

Moore C, Detert JR, Treviño LK, Baker VL, Mayer DM. Why employees do bad things: moral disengagement and unethical organizational behavior. Pers Psychol. 2012;65:1–48.

Nida-Rümelin J. Economic rationality and practical reason. Dordrecht: Springer; 1997.

Palazzo G, Krings F, Hoffrage U. Ethical blindness. J Bus Ethics. 2012;109(3):323–38.

Potemkowski A. Neurobiology of decision making: methodology in decision-making research. Neuroanatomical and neurobiochemical fundamentals. In: Nermend, K. and Latuszynska M, editors. Neuroeconomic and behavioral aspects of decision making. Springer Proceedings in Business and Economics; 2017. p. 3–47.

Robertson DC, Voegtlin C, Maak T. Business ethics: the promise of neuroscience. J Bus Ethics. 2017;144(4):679–97.

Saigle V, Dubljević V, Racine E. The impact of a landmark neuroscience study on free will: a qualitative analysis of articles using libet and colleagues' methods. AJOB Neurosci. 2018;9 (1):29–41.

Salvador R, Folger RG. Business ethics and the brain. Cambridge: Cambridge University Press; 2009.

Sanfey AG. Expectations and social decision-making: biasing effects of prior knowledge on Ultimatum responses. Mind Soc. 2009;8:93–107.

Sanfey AG, Rilling JK, Aronson JA, Nystrom LE, Cohen JD. The neural basis of economic decision making in the Ultimatum Game. Science. 2003;300(5626):1755–8.

Schwartz MS. Ethical decision-making theory: an integrated approach. J Bus Ethics. 2016;139 (4):755–76.

Sen A. Rational fools: a critique of the behavioral foundations of economic theory. Philos Public Aff. 1977;6(4):317–44.

Sen A. Consequential evaluation and practical reason. J Philos. 1997;9(2000):477–502.

Sonenshein S. The role of construction, intuition, and justification in responding to ethical issues at work: the sensemaking-intuition model. Acad Manag Rev. 2007;32(4):1022–40.

Steinmann H, Löhr A. Grundlagen der unternehmensethik. Stuttgart: Schaffer-Poeschel; 1994.

Story GW, Vlaev I, Metcalfe RD, Crockett MJ, Kurth-Nelson Z, Darzi A, Donan RJ. Social redistribution of pain and money. Sci Rep. 2015;5:1–19.

Tenbrunsel AE, Messick DE. Ethical fading: the role of self-deception in unethical behavior. Soc Justice Res. 2004;17(2):223–36.

Tenbrunsel AE, Smith-Crowe K, Umphress EE. Building houses on rocks: the role of the ethical infrastructure in organizations. Soc Justice Res. 2003;16(3):285–307.

Thiel CE, Bagdasarov Z, Harkrider L, Johnson JF, Mumford MD. Leader ethical decision-making in organizations: strategies for sensemaking. J Bus Ethics. 2012;107(1):49–64.

Tomasello M. Why we cooperate. Cambridge, MA: Massachusetts Institute of Technology; 2009.

Trevino LK. Ethical decision making in organizations: a person-situation interactionist model. Acad Manag Rev. 1986;11(3):601–17.

Tversky A, Kahneman D. Judgment under uncertainty: heuristics and biases. Util Probab Hum Decis Mak. 1975;185(4157):141–62.

Walach H, Schmidt S, Jonas BW. Neuroscience, consciousness and spirituality [Internet]. New York: Springer Science and Business; 2011. https://doi.org/10.1007/978-94-007-2079-4.

Werhane PH, Hartman LP, Archer C, Englehardt EE, Pritchard MS. Obstacles to ethical decision-making: mental models, Milgram and the problem of obedience. Cambridge: Cambridge University Press; 2010.

Woiceshyn J. A model for ethical decision making in business: reasoning, intuition, and rational moral principles. J Bus Ethics. 2011;104(3):311–23.

Woodward J. Emotion versus cognition in moral decision-making: a dubious dichotomy. In: Liao SM, editor. Moral brains. Oxford: Oxford University Press; 2016.

Zimbardo P. The Lucifer effect: understanding how good people turn evil. New York: Random House; 2007.

The Social Neuroscience of Empathy and Its Implication for Business Ethics 12

Joé T. Martineau, Jean Decety, and Eric Racine

Contents

Part of this work was realized when J. T. Martineau was a Postdoctoral Fellow at the Montreal Clinical Research Institute, Montreal, QC, Canada.

J. T. Martineau (✉)
Department of Management, HEC Montréal, Montreal, QC, Canada
e-mail: joe.trempe-martineau@hec.ca

J. Decety
Department of Psychology, The University of Chicago, Chicago, IL, USA

Department of Psychiatry and Behavioral Neuroscience, The University of Chicago, Chicago, IL, USA

E. Racine
Montreal Clinical Research Institute, Montreal, QC, Canada

J. T. Martineau, E. Racine (eds.), *Organizational Neuroethics*, Advances in Neuroethics, https://doi.org/10.1007/978-3-030-27177-0_12

Abstract
The neuroscience of empathy is an active area of research that has led to fundamental contributions to our understanding of its cognitive mechanisms and neural circuitry. However, these contributions to knowledge do not constitute a new "brain-based ethics" or the like. Rather, we argue that, situated within an interdisciplinary neuroethics framework inspired by philosophical pragmatism, cognitive neuroscience approaches enrich our understanding of empathy and its role in organizational and business contexts. Henceforth, in this chapter, contemporary neuroscience joins a long legacy of ethics scholarship on the role and possible contributions of empathy in business ethics. First, we review and synthetize knowledge on emotional and cognitive empathy, on empathic concern, and on the role of the different facets of empathy in motivating prosocial behavior and ethical decision-making, with a focus on the insights offered by recent research on the neuroscience of morality. Second, we discuss the contributions of the neuroscience of empathy and ethical decision-making to business ethics and propose five levels of significance: (1) empathy allows for a more informed decision-making in contexts of complexity; (2) empathy is a skill of ethical managers; (3) empathy is a prerequisite to moral engagement; (4) empathy is a rampart against a cold, impersonal, and dehumanized business; and (5) empathy is an essential aspect of efficient organizations. Implications for practice and education are also discussed.

12.1 Introduction

Empathy has for a long time, at least since the eighteenth century, been seen as centrally important in relation to our capacity to gain a grasp of the content of other people's minds and predict and explain what they will think, feel, and do and in relation to our capacity to respond to others ethically. During the past decades, the study of empathy has moved away from mere philosophical discussions. It has become a dynamic and fruitful interdisciplinary area of academic research and applications between the social, psychological, and biological sciences. This has led to new knowledge, which has moved the field forward by contributing to our understanding of the biological and cognitive mechanisms underlying this capacity, how it develops in children, how individual dispositions and situational variables interact in the experience of empathy and outcomes such as helping and caring, as well as its limits (Coplan 2011). More recently, the extent to which empathy impacts social decision-making and moral judgment has led to controversial debates (Decety and Cowell 2015a; Bloom 2016).

These advances may give the impression that a "universal empirical ethics" or a "brain-based ethics" (Gazzaniga 2005) could be generated from such contributions. However, in and of themselves, these contributions hardly constitute a simple foundation for a "brain-based ethics" or the like. The notion of empathy is the object of multiple perspectives. Many authors, from research fields as diverse as psychology, psychiatry, sociology, philosophy, and neuroscience, have studied empathy, offering multiple definitions and lines of inquiry. In general terms, empathy is commonly understood as "the vicarious experiencing of an emotion that is congruent with, but not necessarily identical to the emotion of another individual" (Barnett 1987: 146). However, detailed definitions and views on empathy are far from consensual. Debates still rage between the affective or cognitive nature of empathy, its natural or social genesis, and its influence on altruistic or prosocial behavior, among others. Already two centuries ago, philosopher and economist Adam Smith presented empathy as an interaction between biological, instinctual, emotional, and cognitive processes (Smith 2009 [1790]). However, and unfortunately, these different but complementary processes are still often separated and studied in distinct fields (Wilson 1999; Kitcher 2011), with little integration. It is in the spirit of overcoming this fragmentation that we adopt, in this article, an interdisciplinary pragmatic approach to synthesize knowledge and examine the significance of empathy and emotions in ethical decision-making, reflection, and behavior in the context of organizations (Racine 2008; Racine 2010).

We argue that, situated within an interdisciplinary neuroethics framework inspired by philosophical pragmatism, neuroscientific research enriches our understanding of empathy and its role in organizational and business contexts. Indeed, we believe that a series of interdisciplinary convergences found in the literature allows us to increase and deepen our understanding of empathy and its implication for business ethics. This approach corresponds to the naturalist philosophy of John Dewey (1922), who insisted on the significance of an interdisciplinary and empirical study of morality. Our goal here is to highlight the potential of an interdisciplinary reflection on empathy and emotions, including evidence from the field of neuroscience, to inform business ethics. In doing so, this chapter aims at building bridges between social neuroscience and (business) ethics, thus contributing to the fields of neuroethics (Racine 2010).

In the first part of this chapter, we review neuroscience research on different components of empathy and the role of empathy in motivating prosocial behavior, based on literature from social neuroscience but also from philosophy and social psychology. We specifically address how neuroscience has enriched previous definitions and understandings of empathy and discuss how empathy, by giving people access to others' emotions, enables them to make more ethical decisions. In the second part, we discuss the significance and implications of these interdisciplinary findings on the role of empathy and emotions in ethical decision-making for the study, practice, and pedagogy of management and business ethics.

12.2 Definitions, Mechanisms, Modulators, and Functions of Empathy

In the last century, the notion of empathy has been intensely explored by philosophers and psychologists, mainly because of its relevance for moral theory. Empathy was viewed as potential source of moral behavior, but also because it has revealed to be an essential aspect of interpersonal understanding. More recently, research in social neuroscience offers new insights for the understanding of this complex phenomenon, mostly through neuropsychological studies of psychopaths or of individuals suffering from brain damage and through studies using functional neuroimaging techniques with healthy participants. Contemporary theoretical and empirical work from evolutionary biology, developmental psychology, and social neuroscience converges to view empathy as a multidimensional construct reflecting a natural capacity to share and understand the subjective states of others and comprising emotional (affect-sharing *with* another), cognitive (understanding the other's subjective state from his/her point of view), and motivational (feeling concerned *for* another) facets (Decety and Jackson 2004; Leiberg and Anders 2006; Decety 2015). These facets or components operate by way of automatic and controlled processes and interact with one another. They are intertwined, and yet they can be dissociated, as they rely on partially separable information-processing systems.

With research stemming from diverse scientific disciplines, there exist multiple views and definitions of empathy (Batson 2009). We will focus hereafter on the neurophysiological underpinnings of affective and cognitive empathy, their modulators, and how they relate to morality.

12.2.1 Affective Empathy

Since empathy is often seen as an umbrella term (Batson 2009; Decety and Cowell 2014b), it is important to distinguish the affective aspect of empathy from other connected phenomena like emotional contagion, emotion sharing, and sympathy. Affective empathy, also referred to as affect-sharing, accounts for the vicarious experience of another's emotions or the experience of "coming to feel as another person feels" (Batson 2009: 5). As such, the affective component of empathy "provides us with an experiential access to other minds" (Zahavi and Overgaard 2012: 10). Emotional contagion slightly differs from affective empathy as it is an "automatic transmission of emotions between individuals" (Hatfield et al. 1994: 11) that mostly involves one's automatic mimicry of the posture, facial expression, voice, and movements of other people that allows for emotions to converge (Hatfield et al. 1993), like in the case of the spread of a mass panic, without individuals actually knowing and understanding the origin of their fear. In this case, there is an isomorphic emotional response between individuals that can also be seen in affective empathy, but this phenomenon does not meet the "self-other distinction" that occurs in empathic response (Preston and de Waal 2002; de Vignemont and Singer 2006) (i.e., the individual is not necessarily aware that his/her affective state is generated by the affective state of another person). To put it simply, in empathic response, the individual is aware that the distress he/she might feel when witnessing the distress of his/her friend is not his/her distress, but his/her friend's. Thus, the locus of empathy is "other-centered," while emotional contagion is "self-centered" (de Vignemont 2009). Second, emotion sharing also differs from empathy and relates to the simultaneous sharing of an emotion, such as when two people experience the same emotions when watching a movie together, or to the human propensity to recount emotions to others. Third, sympathy, also often referred to as "empathic concern," differs from affective empathy, as it is more closely associated with compassion—feelings of sorrow or concern for another person or "feeling for" another person who is, for example, in pain or suffering (Batson 2009; Eisenberg and Eggum 2009)—than it is to "feeling with" the other. There is clear neuroscience evidence that affective empathy and empathic concern are supported by distinct neural systems (Ashar et al. 2017). Although affective empathy may lead empathic concern, which is a more active state, it is not necessarily always the case. However, when empathy gives way to empathic concern, it is more likely to make people act on their emotions and to be the source of altruistic behavior.

The study of the neurophysiological bases of empathy historically began with the study of individuals suffering from mental disorders or brain damage who showed different levels of dysfunction in empathic response. A variety of mental disorders and disability can be associated with a deficiency in empathy, such as autism, narcissistic personality disorder, psychopathy, antisocial personality disorder, and alexithymia (Decety and Moriguchi 2007). For example, this last disorder, alexithymia, is a personality trait characterized by a deficit in the cognitive processing of emotions (i.e., the inability for a person to recognize and describe his/her own emotions) (Taylor et al. 1991). This trait, present in approximately 10% of the population (Salminen et al. 1999), has been revealed as a modulator of empathy in the sense that the more the individuals have difficulty identifying their own emotions, the less they can empathize with others (Bird et al. 2010). Moreover, lesion studies of patients suffering from damage within the right somatosensory cortices found that these patients had impaired ability to recognize others' emotions when viewing their faces (Adolphs et al. 2000). Patients suffering from this type of brain damage also show impaired emotional response, a phenomenon witnessed in the famous case of Phineas Gage (Damasio et al. 1994). Thus, it seems that in order to recognize emotions in others, individuals have to reconstruct a similar emotional state in their own brain, as if they were experiencing it themselves, which is impossible for these patients. This hypothesis is supported by Preston and de Waal (2002), who suggested, in their perception-action model (PAM) of empathy, that in order to apprehend another person's internal emotional state, the observer must represent in his/her own brain a corresponding existing representation of the state acquired in past experiences, which "automatically primes or generates the associated autonomic and somatic responses" (Preston and de Waal 2002: 4). This view is also coherent with Damasio's somatic marker hypothesis (Damasio 2005). At the neurobiological level, it is thought that the process of motor empathy (mimicry) would imply mirror neurons, a class of cells with sensorimotor properties discovered in the mid-1990s in the ventral premotor cortex of monkeys. However, several studies have challenged the actual involvement of mirror neurons in empathy and sympathy (Decety 2010; Fan et al. 2011) and argued that while these neurons play a role in motor learning, it is not clear that they are implicated in emotion processing.

Studies on the neurochemical mechanisms of empathy add another level in the understanding of this capacity. In particular, the neuropeptide oxytocin, known for a long time to facilitate the development of mother-infant bonding and caring, has been identified as playing a role in social cognition including trust, empathy, and generosity (Beetz et al. 2012; Ebert and Brüne 2017). Accumulating evidence suggests that oxytocin can enhance empathy although it is unclear which specific behavioral and neural aspects are influenced (Geng et al. 2018). In addition, genetic variation in the oxytocin receptor (OXTR) has been associated with individual differences in the expression of various social behaviors including empathic concern (Smith et al. 2013).

In the past decade, a multitude of studies using neuroimaging methods such as functional MRI (fMRI) and electroencephalography (EEG) have provided new insights into the neural mechanisms involved in empathy. This research provides some support for the hypothesis of shared neural representations involved in the first-hand experience of an emotion and empathic response to a similar emotion. This seems particularly the case for pain empathy. It has been demonstrated that specific brain areas, namely, the anterior insula and the anterior cingulate cortex, are activated both when an individual is experiencing pain and when this individual is witnessing another person in pain (Lamm et al. 2011). Moreover, some studies show that the somatosensory cortex is also activated when participants are presented with strong and explicit visual stimuli of infliction of pain to other persons or body parts (Bufalari et al. 2007; Lamm et al. 2010; Gonzalez-Liencres et al. 2013). EEG studies have documented specific early and late neural components in response to the pain or the distress of others, which can be used to distinguish between affective empathy and empathic concern (Decety et al. 2015).

As for empathy involving other emotions, such as anger, anxiety, disgust, fear, or happiness, an fMRI meta-analysis identified a consistent activation of the dorsal anterior cingulate cortex, the anterior medial cingulate cortex, and the anterior insula, suggesting that these brain areas constitute a core neural network for empathy (Fan et al. 2011). However, a distinction must be made between the different routes to the generation of empathic responses—either as a response to the perception of a direct stimulus or coming from a more abstract inferential path (Engen and Singer 2013)—and their corresponding neural substrates. Empathy involves not only an affective experience of another's emotion when witnessing this other person's emotional state but also a form of recognition and understanding of this emotional state (Decety and Moriguchi 2007), a phenomenon to which we will refer to as cognitive empathy.

12.2.2 Cognitive Empathy

In the literature, the cognitive aspect of empathy is seen as the capacity to adopt the subjective perspective of another person. Cognitive empathy partly overlaps with the concept of theory of mind, which refers to the ability to attribute mental states, thoughts, and desires to others and oneself, a capacity of humans that is part of our social cognition and that develops in children at a young age (Leslie et al. 2004).

In contrast with affective empathy, which is considered a bottom-up process that occurs spontaneously and without awareness, cognitive empathy is a top-down process that implies executive functioning (Decety and Moriguchi 2007; Preston et al. 2007) and is generally defined as the ability to consciously put oneself into the mind of another person to understand what he/she is thinking or feeling (Decety and Yoder 2016). Bryant (1987) captured this difference between affective and cognitive empathy, when she suggested that "social perspective taking entails cognitive understanding of the feelings and motives of others and, as such, is an instrumental skill. Empathy, on the other hand, entails emotional responsiveness to the feelings experienced by others and, as such, is an expressive experience" (Bryant 1987: 246). However different, these two processes are intertwined, and it is their dual action that accounts for the full experience of empathy in humans.

Neuroscience also provides evidence for the distinction between the affective and cognitive facets of empathy. As explained before, a response to direct visual stimuli, such as the witnessing of another person in pain, results in activation of the core neural network, including the anterior cingulate cortex and insula. In contrast, when the direct stimulation is missing, individuals can create representations of another person's potential mental state, for example, in the case of somebody being told that a friend has lost her husband in a car accident. In this case, the generation of an empathic emotional response toward this friend is made possible by the use of perspective taking, knowledge of the individual in question, and theory of mind: all processes that are associated with the cognitive dimension of empathy. Consequently, the activation of the core neural network of empathy identified earlier is consequent to the activation of neural networks involved in theory of mind, mentalizing, and perspective taking and valuation (Ruby and Decety 2003; Amodio and Frith 2006; Jenkins and Mitchell 2011). Importantly, the ventromedial prefrontal cortex is a critical hub for empathic concern and social decision-making (Decety and Cowell 2014b; Yoder and Decety 2018). This distinction between the emotional and cognitive aspects of empathy is one of the significant contributions of the neuroscience of empathy, and it has important implications in the field of ethics, as we will discuss later.

12.2.3 Modulators of Empathy

The degree to which individuals empathize with others is modulated by social and interpersonal factors. Research in neuroscience and social psychology has documented that empathy and prosocial behaviors are modulated by the degree of affiliation, affective link, proximity, and similarity between individuals. People demonstrate more empathy toward kin (Cheng et al. 2010) group members than toward unaffiliated others (Hein et al. 2010; Echols and Correll 2012). Apart from the nature of the relationship between the empathizer and the other person, other modulators of empathy have been identified, such as characteristics of the emotion (e.g., valence, intensity, saliency), of the empathizer (e.g., mood, personality, gender, age, past experience), of the context, and of the goals, beliefs, and values we bear and possibly share with this other person. Hence, empathy is shaped by social context (Decety 2015). Also, as presented before, at the neurobiological level, oxytocin and its gene receptor have been identified as a modulator of affective empathy (Smith et al. 2013). It has also been suggested that other neurotransmitters, such as dopamine and serotonin, are neurochemical mediators of social cognition or theory of mind [for a review of these studies, see Decety et al. (2012) and Gonzalez-Liencres et al. (2013)].

12.2.4 Functions of Empathy

To sum up our review so far, we can distinguish between two forms of empathy: one that appears spontaneously and automatically and one that is more intellectual and that can then be more easily cultivated. The spontaneous form of empathy is a capacity that develops in humans at a young age to be able to put themselves in the place of others, directly or indirectly, by the power of imagination, cognitive, emotional, and neurobiological processes, to feel what another might feel if he/she were in his/her place. In this view, which encompasses both affective and cognitive aspects described before, empathy is a form of *knowledge* (Batson 2009). The cultivated form of empathy is an extension of the spontaneous one that becomes a deliberate modus operandi, an approach to social relations characterized by a voluntary openness to other's perspectives, emotions, and feelings. Here, empathy, mostly the cognitive aspect of it, is viewed as an action or attitude.

Empathy is believed to play an essential role in morality. Indeed, "empathy-related processes are thought to motivate prosocial behavior, inhibit aggression, and provide the foundation for care-based morality" (Decety and Cowell 2015b: 279). But we must distinguish between the different facets of empathy and their involvement in motivating prosocial behavior (Decety and Cowell 2014a). Indeed, there is a strong consensus between neuroscientists that empathic concern, which corresponds to the motivational component of empathy that urges us to care for others, plays an important role in motivating prosocial or altruistic behavior (Batson 2012; Decety and Cowell 2014b; Decety 2015). For example, a large sample study has consistently shown that low levels of empathic concern are a predictor of utilitarian moral judgment in personal or impersonal moral dilemmas (Gleichgerrcht and Young 2013). Both affective and cognitive empathy can lead to empathic concern; however, there is a controversy as to whether the former may lead to distress rather than to caring for others (FeldmanHall et al. 2015; Ashar et al. 2017). Also, a study by Decety and Yoder (2016) shows that it is the cognitive and motivational components of empathy, and not affective empathy, that are associated with an increased motivation for justice.

In general, and in the context of organizations, we believe empathy bears ethical significance because both knowledge of and openness to others' emotions are essential aspects of ethical decision-making and ethical management. Indeed, empathic processes give us access to others' emotions, which are in turn valuable inputs in our own decision-making process and which play an important role in motivating our ethical behavior. This central role of emotions in morality has been attested by authors from diverse fields. Precisely, authors have shown that emotions act as moral intuitions that contribute to the evaluation and appreciation of a situation (Damasio 2005). Also, neuroscience research has demonstrated that emotions are intrinsic and essential aspects of decision-making (Damasio 2001; Moll and de Oliveira-Souza 2007), a position also supported by authors in management and business ethics (Ashforth and Humphrey 1995; Seo and Barrett 2007). Therefore, the emotions of individuals who are affected by a specific situation can be accessed through empathy, transformed into knowledge, and integrated into new intuitions, which in turn enrich ethical reflection, decision-making, and behavior on the part of the empathizer.

In the next part, we discuss the organizational implications of the findings about empathy presented in the first part of this chapter. We suggest five levels of significance for business ethics.

12.3 Implications of the Neuroscience of Empathy for Business Ethics

In the first part of this chapter, we presented the contributions of neuroscience, among other disciplines, to the understanding of empathy and its role in moral judgment, ethical decision-making, and behavior. Globally, the neuroscience of empathy has contributed to (1) distinguishing between the emotional and cognitive aspects of empathy, their underlying neural processes (Eres et al. 2015), and their modulators; (2) rejecting the view of ethics as only rational by confirming the importance of affect in morality; and (3) revisiting the traditional distinction between reason and emotion in ethical decision-making. Additionally, with regard to the motivational impact of empathy on behavior, although research on this complex topic is ongoing, neuroscience now "lends strong support to the notion that emotion reactivity in general, particularly empathic concern, play a pivotal role in guiding prosocial behavior" (Decety 2015: 4).

Either as a form of knowledge or as a disposition, both the affective and cognitive aspects of empathy provide access to the emotions and minds of others. Given the central role and importance of emotions and beliefs in ethical reflection, decision-making, and behavior, empathy is consequently of major significance in business contexts, both at the individual and organizational levels. In this next part, we discuss how the findings presented in the first part of this chapter echo in the field of business ethics and identify five of their contributions in organizations (Fig. 12.1). We also discuss the limitations of empathy in organizations and the implications of our findings for the education of managers and future managers.

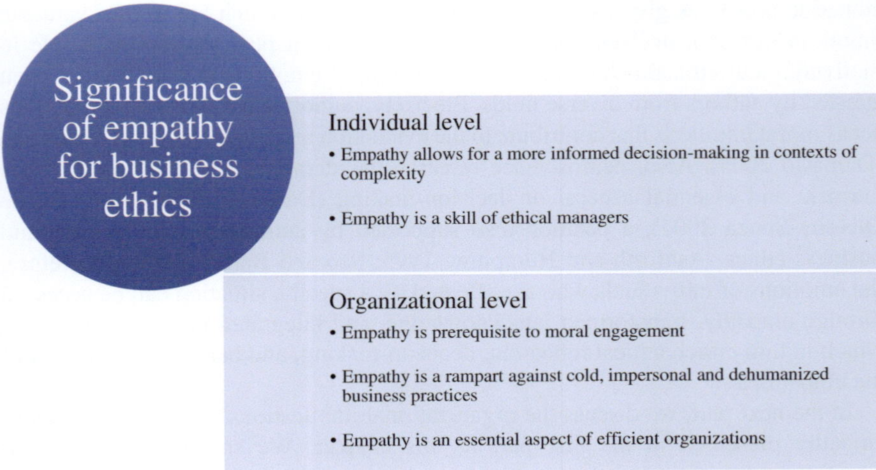

Fig. 12.1 Significance of empathy for business ethics

12.3.1 Empathy Allows for More Informed and Grounded Decision-Making in Contexts of Complexity

Empathy is an important mode of understanding the subjective states of others. Emotions are themselves intuitions that are essential to guide decision-making and enrich reflection about ethical situations. Thus, empathy, as a positive emotion, plays an important role in the genesis of grounded moral judgments (Pizarro 2000) and ethical decision-making. First, others' emotions obtained through empathic processes can serve as triggers that signal the presence of an ethical dilemma (Gaudine and Thorne 2001) or a morally inappropriate decision (Dietz and Kleinlogel 2014). In other words, empathy acts as a moral marker that indicates the presence of a morally important event, and it "stimulates the individual to think about the moral principles that may hold appropriate jurisdiction over the current situation" (Pizarro 2000: 360).

Second, both affective and cognitive empathy enrich our interpretation of situations because they represent additional sources of information. By giving us an "inside view" of the affective lives of others, "[emotional] empathy improves or deepens [our] understanding of others' emotions that we get from other sources" (Zahavi and Overgaard 2012: 15). Given that ethical issues and dilemmas at work and in other contexts are often emotionally charged, better understanding of the emotions at stake is an important factor in their analysis. Also, as discussed earlier, research in neuroscience has suggested that cognitive empathy is especially important to allow individuals to create mental representations of the mental state or emotions of others, even when not in direct contact with the individuals potentially experiencing these emotions. Cognitive empathy is therefore essential in the evaluation of possible consequences of decisions for vulnerable stakeholders, which is one aim of stakeholder management philosophy (Donaldson and Preston 1995). Indeed, the more people take the perspectives of others into account in their analysis of a situation, "the more likely they are to consider their responsibility towards others and form highly ethical intentions" (Mencl and May 2009: 218).

Third, empathy works against fragmentation in organizations. According to Pavlovich and Krahnke (2012), empathy presupposes both an interest toward others that supports the transcendence of one's self-interest and an openness to other paradigms that foster enlighted relationships and connectedness between individuals. In pluralistic contexts, such a disposition is a harbinger of successful resolution of tensions surrounding ethical principles and human values. This openness to other paradigms gives managers a greater understanding of contexts and makes them better equipped to analyze complex situations. Indeed, according to Barnard (1938), "brains without minds seem a futile unbalance." He suggested that the complexity of organizational dynamics "need the corrective of the feeling mind that senses the end result, the net balance, the interest of the all and of the spirit that perceiving the concrete parts encompasses also the intangibles of the whole" (Barnard 1938: 322).

In sum, contextualized openness, and sensitivity to the emotions of others allow for a richer interpretation of complex situations. Thus, empathy contributes to better-informed moral judgments and ethical decision-making in organizational settings.

12.3.2 Empathy Is a Skill of Ethical Managers

From the perspective of management skills and competencies, empathy is a distinctive asset not only used to properly understand complex human and social situations but also to find satisfactory resolution to a range of issues and dilemmas in practice. We suggest that in organizational and business contexts, the nurturing and culture of empathy could be counted as a professional virtue. Others have taken a similar stance, highlighting empathy as an important asset of business managers (Lurie 2004), as a leadership skill (Kellett et al. 2006; Holt and Marques 2012), or as a safeguard for ethical managerial action toward the organization and its stakeholders (Dietz and Kleinlogel 2014). Empathy contributes to successful social interactions by allowing individuals to predict and understand other's behaviors and to react accordingly (Singer and Fehr 2005; Engen and Singer 2013). Additionally, studies have demonstrated that mimicking of another person's gestures, postures, mannerisms, and other motor movements increases rapport and prosocial behavior (van Baaren et al. 2004; Chartrand and Lakin 2013). This implies that the automatic mimicry that often accompanies empathy can lead to "greater liking, helping behavior, closeness and interdependence with others, accuracy in understanding emotions, and reduction in prejudice" from the persons who are being mimicked (Chartrand and Lakin 2013: 300). This is of importance in management, as it has been shown that this aspect of empathy improves the smoothness of social interaction (Decety and Moriguchi 2007). Empathy is then useful from a self-interested point of view, although we can argue that this capacity of empathizing also undermines purely self-interested decision-making and leads to more other-regarding behavior.

Studies on emotional intelligence tend to converge with this understanding of empathy as a managerial skill. Emotional intelligence is globally defined as the capacity to reason about one's own and others' emotions and of emotions to enhance thinking (Mayer and Salovey 1997). More specifically, emotional intelligence embraces four dimensions that can be associated with different aspects of empathy discussed before: (1) the capacity to perceive and recognize emotion in the facial and postural expressions of others; (2) the capacity of emotions to assist in thinking; (3) the capacity to analyze emotions, to understand their evolution over time and their outcomes; and (4) the management of emotions in the context of the individual's goals, self-knowledge, and social awareness (Mayer et al. 2004). Indeed, affective empathy has been identified as a key component of emotional intelligence (Goleman 1995; Law et al. 2004). Emotional intelligence has been showed to be a predictor of different prosocial and positive behaviors, including positive leadership and other organizational behaviors (for a review, see Mayer et al. 2004), and an essential component of ethical judgment in managers (Angelidis and Ibrahim 2011). In that light, the ability to identify with others, to understand their emotions, and to empathize with them can be seen, again, as a skill of ethical managers.

Conversely, cutting oneself off from affective empathy, creating a barrier between a person and their own emotions as well as the emotions of others, could mean the loss of precious intuitions. This could lead to both a less informed analysis of a situation and less informed decision-making. In the context of ethical issues, it could also mean the loss of emotional intuitions that could generate impulse for altruistic or moral behavior, instead leading to moral silence or inaction (Bird 1996), a factor that has been identified in the study of various ethical scandals that have emerged in the past decades, including Enron (Trinkaus and Giacalone 2005).

This is not to say that managers have to reject logical analysis of situations and rely solely on emotions and intuitions. The difficulty of management lies differently in the ability to exert rationality while still paying attention to the valuable emotional and intuitive knowledge obtained through one's own emotions and through others' emotions via affective or cognitive empathy. In sum, as Barnard said, "to rely upon 'feeling', to give weight to first impressions, to reject logical conclusions and meticulous analysis in favor of an embracing sense of the whole, involves an inconsistency in attitudes. It means developing the artistic principle in the use of the mind, attaining proportion between speed and caution, between broad outlines and fineness of detail, between solidity and flexibility" (Barnard 1938: 322).

12.3.3 Empathy Is Prerequisite to Moral Engagement

Bandura studied the different psychosocial disengagement mechanisms and practices used by individuals and social systems to inhibit individual's moral self-sanctions and control and allow for the perpetration of inhumanities (Bandura 1999).

One of these mechanisms, the disregard or distortion of consequences of inhumane conducts, is facilitated by physical or hierarchical distance between the offender and the victims of his/her conduct. Inversely, when people can see and hear the harm and suffering they are causing, they feel a vicarious distress that serves as self-restrainer for their actions (Bandura 1999). This process, to which we referred earlier as affective empathy, can thus be considered a factor of moral engagement. However, as stated earlier, more research is needed on the relationship between affective empathy and morality, as it still debated (FeldmanHall et al. 2015; Ashar et al. 2017).

The neuroscience of empathy has shown that affective empathy is modulated by physical proximity, among other factors, in the sense that the more an empathic connection can be established in a direct and sensory way between people, the more they will be likely to engage morally, an idea also put forward by Bandura (1999). Again, as discussed before, affective empathy is also increased with social, cultural, or psychological proximity, meaning that people care and empathize more with individuals they can identify with. In management literature, proximity, in its different dimensions, was identified as a factor of moral intensity in Jones' (1991) contingent-issue model of ethical decision-making. However, when this proximity is not present, which is often the case in managerial decisions with potentially negative consequences for remote stakeholders, neuroscience research has revealed that voluntary cognitive empathic engagement and perspective taking can take over and palliate for the reduced affective response caused by social, psychological, or physical distance (Decety and Cowell 2015b). Cognitive empathy can thus aid managers in forming a more ethical decision with responsibility toward these distant stakeholders. This corresponds in some way to the "names-and-faces" stakeholder management approaches proposed by McVea and Freeman (2005), which can be considered as a way to foster proximity, identification, and empathic responses from managers toward stakeholders. Therefore, because it can help people to overcome the boundaries of affective empathy, we also consider cognitive empathy as a factor of moral engagement.

Authors have also suggested that emotions, including empathy, can become collective (Smith et al. 2007; Muller et al. 2014) and that emotions can define groups as more than just a collection of individuals. Empathy can influence and motivate collective and managerial decisions and prosocial behaviors, such as corporate philanthropy (Muller et al. 2014), and thus become a factor in collective moral engagement.

Still, we cannot deny that there exists a popular perception, and plenty of actual cases, of business people being morally disengaged and that empathy might not be the first personality trait associated with business people. Indeed, there is substantial literature showing that business students, especially students in economy or finance majors, are less cooperatives and more likely to cheat than students from other fields (for a review, see Brown et al. 2010). A study by Brown et al. (2010) about the impact of empathetic or narcissist personality traits on ethical behavior in business students supported the idea that empathy has a significant positive effect on ethical decision-making. Interestingly, finance majors showed significantly less empathic and more narcissistic responses than business students in other majors. For the authors, this result was not surprising, considering the pecuniary values emphasized in the finance curriculum, including the need to commodify all aspects of a business, including human resources, and the focus on rational and quantitative analysis of situations that both may lead students to believe that emotions and empathic concern are not part of the equation. It might also be that finance studies attract, from the beginning, people with lower levels of empathy, a problem that the curriculum does nothing to correct. And then the story repeats itself and feeds a certain culture of moral disengagement that persists in this sector. However, the fostering of empathy among business students might encourage them to engage morally and could contribute to fight against a dehumanization of business practices.

12.3.4 Empathy as a Rampart Against Cold, Impersonal, and Dehumanized Business Practices

Business ethics found its genesis in the reaction (the outrage) to numerous scandals, reflected in a profound need to change the way we do business. Indeed, in the last decades, the emergence of large financial scandals with ethical connotations, notably the cases of Enron or Arthur Andersen, as well as the profound banking and financial crisis that started in 2008, brought about new considerations for the public in general and in the fields of management and organization studies in particular. Corporations are now often depicted as arenas, as battlefields of cutthroat competition fed by a mean corporate culture, focused on competition and on maximizing performance regardless of all social costs. As discussed before, business people are more often than not associated with a stereotype of cold, impersonal, and selfish individuals enriching themselves at the expense of others. Unfortunately, the studies of ethical scandals that affected several corporations already taught us that this stereotype, depicted in numerous movies and books, is not so far from a certain reality.

Therefore, the presence and consequences of what we have come to know as "corporate psychopaths" have become topics of interest and research. Corporate psychopaths have been described as superficially charming with a grandiose sense of self-worth, manipulative, pathological liars, calculating, cold, and ruthless, showing no remorse or sense of responsibility for their actions, and lacking empathy (Hare 1994; Boddy 2014). Contrary to previous beliefs about their lack of success in work environments, it appears that psychopaths can be very successful in different types of work settings and can thrive in the business world, bringing negative consequences on their colleges and the organizational culture (Babiak 1995; Clarke 2005; Babiak and Hare 2006). Although it might not be possible to change these individuals, organizations that want to improve their ethical culture might act in recognizing the presence and the impacts of corporate psychopaths and avoid putting them in positions of authority and influence. They might also include empathic assessment in their hiring process to try to select, at this time, individuals who exhibit more developed empathic skills.

Empathy on the part of managers and employees contributes to a more ethical organizational culture. Indeed, neuroscience literature supports the fact that higher levels of empathy lead to more cooperation between individuals, which can then lead to more reciprocity in relationships. These, in turn, reinforce empathic responses between people, creating a positive feedback loop (Gonzalez-Liencres et al. 2013) that feeds the culture from which it emerged. Also, according to Lurie (2004), emotions make more humane managers and better leaders and contribute to humanizing business in general. Globally, empathy allows individual to understand why it is wrong to harm others (Decety 2015). Hence, in light of the context described before, we believe empathy is a crucial ingredient in re-humanizing the world of business and organizations and in fighting against this culture of selfishness by engaging with others in more cooperative behaviors (Mintzberg et al. 2002).

12.3.5 Empathy as an Essential Aspect of Efficient Organizations

Although distinct from the ethical underpinning for the justification of business ethics, increased empathy could also very well be a component of successful organizations more globally speaking. First, empathy is a facilitator of positive and effective teamwork (Kayworth and Leidner 2001). Indeed, neuroscience research has shown that higher levels of empathy lead to more cooperation and reciprocity between individuals, facilitate communication, improve group identification and cohesion, and promote behaviors that contribute to the fitness of the group as a whole (Gonzalez-Liencres et al. 2013). Empathy can also help individuals to recognize people who are marginalized in the group and act upon their better integration or to detect deceit from other group members (Echols and Correll 2012).

Also, this cooperation facilitated by empathy does not seem to be restricted only to the internal relations of organizations. Dutton et al. (2006) studied the response of organizations toward victims of social and environmental crisis and highlighted the important role of empathy in this process. According to the authors, emotions, such as empathic concern, are constructive forces that contribute to the phenomenon of compassion organizing by facilitating the activation and mobilization of resources in times of crisis.

Empathy in leaders could also lead to increased leadership effectiveness. Humphrey (2013) proposed that entrepreneurs with greater empathic skills would be more successful at motivating employees, would generate less stress for their employees and help them cope with stressful situations, and would be more attuned to their customers' wants, needs, and satisfaction. Moreover, a study on the emotional intelligence of leaders showed that top executives who possessed higher levels of empathy, among other things, were more likely to yield high profits for their company (Stein et al. 2009).

In the last years, there has also been a proliferation of management books about the effectiveness of empathy, depicting empathy as a source of competitive advantage, as a key ingredient of business prosperity, or as a management philosophy, for example. Of course, these popular books are not based on empirical research, but they are still an indicator of the increased interest in the subject of empathy in the management community.

12.3.6 Limitations of Empathy in Organizations

Empathy, in the context of ethics in general and business ethics, is globally considered for its positive influence on decision-making (Decety and Cowell 2015a). However, it is important to acknowledge that there are some possibly negative aspects that need to be addressed, as they also correspond to potential negative implications for business ethics. First, empathy felt for a person can lead people to violate moral standards of fairness (Batson et al. 1995) or the utilitarian principle of acting for the greater good of the greatest number (Batson et al. 1999). As shown by research in psychology and neuroscience, it can result in the partial treatment of people to the benefit of the one with whom the decision-maker have empathize (Decety and Cowell 2014b). In the context of business, this could lead to favoritism, benefitting group members, and nepotism. Also, empathy can lead to a paternalistic response to the person in need, without it being the real help they would need (Batson 2012).

Moreover, given the influence of certain modulators of empathy, as explained earlier, some empathic responses may lead to negative feelings, such as envy or even a malicious joy (*schadenfreude*), when witnessing another's misfortune, depending on the social context and/or the previous relation between the empathizer and the other. These negative responses can also enhance competition and non-cooperative behaviors between individuals, which in turn reinforce negative feelings from empathic responses, in a negative feedback loop (Gonzalez-Liencres et al. 2013). In the context of business, highly competitive organizational environments and cultures may thus lead to negative feelings from empathic responses and non-cooperative behaviors, which will again reinforce the competitive culture from which they emerged.

Also, it should be recognized that people can sometimes be selfishly motivated to consciously avoid being empathetically concerned for others in order to protect them from the altruistic motivation in which it could result (Batson 2012). In this case, the automatic emotional empathic response to another's emotion can be voluntarily controlled or inhibited by a person, as much as the empathic disposition reduced to its minimum, like in the case of corporate psychopaths, as discussed earlier. In the same vein, Machiavellian behavior has been vastly studied in the context of organizations, and it is interesting to mention that an fMRI study of empathic responses in managers has linked Machiavellianism with low cognitive empathic capabilities, combined with high automatic affective empathic response (Bagozzi et al. 2013). According to the authors, the increased ability to attune to the emotions of others (especially negative emotions) demonstrated by Machiavellian individuals could facilitate, for them, the manipulation of others.

12.3.7 Implications for Education

In spite of the potential negative outcomes of empathy described above, as presented in the first part of this chapter, it is generally recognized that empathy and emotions obtained through empathic processes play a positive role in morality and are essential aspects of ethical decision-making. That being said, some of the more difficult questions are raised by the eventual proactive teaching and nurturing of empathic dispositions. As stated before, the cognitive dimension of empathy may be considered a skill that can be developed and cultivated through different means. Even the affective aspect of empathy, which we referred to as an automatic process, might also be nurtured. Indeed, neuroscience research has demonstrated that even if empathy and subsequent prosocial behaviors are shaped by social context and that people usually favor in-group members, this tendency is not unalterable and can be overcome (Decety 2015). For example, in an EEG/ERP (electroencephalography) study, Sheng and Han (2012) showed that the racial bias in empathy can be decreased by the use of cognitive strategies such as instructing participants to pay attention to and observe others' feelings of pain, which increased neural responses to pain expressions for other-race individuals. Hence, perspective taking, an aspect of cognitive empathy, can be used as a strategy to elicit empathy, concern for others, and altruistic motivation (Batson et al. 1997; Decety and Cowell 2015b).

Also, recent studies in aging have suggested that neuroplasticity occurs not only in children and adolescents but also in adults and throughout the whole lifespan (Mora et al. 2007; Boyke et al. 2008; Taubert et al. 2010; Schlegel et al. 2012). This means that human brain matter and structures can still develop in adulthood, in response to skill acquisition, learning, and our ever-changing environment. Relying on these new findings, Georgi et al. (2014) have suggested that empathic abilities can be developed in adults through the continuous experience of interpersonal socio-emotional stimulation and challenges.

All in all, it is now generally accepted that empathy can be learned and taught (Feshbach and Feshbach 2009). As Adam Smith had already noted in his time, empathic capacities and dispositions can be stimulated by daily activities, such as sound relationships at work, but also through education and exposure to art, music, or literary works, allowing people to develop their imaginative, emotional, and aesthetic faculties—an idea also supported by contemporary scholars from different domains (Smith 2009 [1790]; Nussbaum 2010; Kidd and Castano 2013; Decety and Cowell 2014b).

In management literature, Gaudine and Thorne (2001) suggested that organizations should encourage their employees to recognize their emotions at work, what they signal, and their possible influence on their decision-making process. Mencl and May (2009) recommended using training programs or educational strategies to encourage managers to recognize different types of information, including emotions, which are part of their ethical decision-making process. They also suggest that managers should be encouraged to get to know their employees and stakeholders more closely, so as to encourage empathic responses toward them. Also, authors suggested that business school education, management development programs, and executive coaching can and should be used to promote empathy in leaders (Mahsud et al. 2010; Holt and Marques 2012).

We believe there is also a need to improve empathic education for business students. This is in line with Zhong (2011) assertion that we might be wrong to focus business education on rational, deliberative thinking. He suggests that economic training might cultivate a culture of "cold-blood" decision-making that overshadows the influence of intuition, which can in turn lead to unethical decisions and behaviors. A similar discussion is taking place in the medical education, where the hidden curriculum, suspected to encourage emotional detachment and cynicism, combined with other factors such as a lack of role models and time pressure, results in an important decline of empathy in third-year medicine students (Hojat et al. 2009), a phenomenon also witnessed in nursing students (Ward et al. 2012). To overcome this problem, Riess et al. (2012) have proposed and tested a neuroscience-informed addition to the curriculum of medical students, consisting of three 60-min empathy training modules integrating knowledge about the neurobiology of emotions and empathy. The authors showed that this brief intervention resulted in improved physician empathy, as rated by patients, and improved ability to decode the emotions contained in subtle facial expressions. This might be an interesting avenue for the enrichment of the curriculum in business schools.

Indeed, Brown et al. (2010) have suggested that business schools should make an effort to include empathy development in their curriculum, not in a theoretical way but by providing students with multiple opportunities to develop ethical traits throughout the whole pedagogical experience. This idea is also supported by Cohen (2012), who insisted that business schools can cultivate empathetic reactions and cooperative outcomes on the part of students through service-learning projects, for example. However, more research and guidance is needed in how exactly business schools should implement empathy development in the curriculum and which pedagogical approaches to use in the classroom (for a review of pedagogical procedures and approaches, see Feshbach and Feshbach 2009). Future advances in cognitive neuroscience may eventually provide an even better understanding of empathy and its specific processes and insights into how to foster empathy among student and managers. Practically, pedagogical experiments such as the one conducted by Riess et al. (2012) still have to be tested in business school contexts. Finally, we also believe a more complete interdisciplinary account of the phenomenon of empathy in education and training, similar to the one adopted in this chapter, might lead to improved ethics and business ethics training, focusing on empathy. This could be, in our opinion, a promising and engaging way to improve ethics in organizations.

12.4 Conclusion

This chapter drew attention to the contributions of the social neuroscience of empathy to ethical reflection, decision-making, and behavior in organizations. Neuroscience evidence help us better understand the complementary aspects of affective and cognitive empathy and their underlying mechanisms and highlight their importance and significance in organizational contexts. Finally, we believe that the deepening of a culture and practice of empathy, both at the individual and organizational levels, can complement and enrich the rationalist perspective prevailing in management and business ethics and its associated pedagogical and practical approaches.

References

Adolphs R, Damasio H, Tranel D, Cooper G, Damasio AR. A role for somatosensory cortices in the visual recognition of emotion as revealed by three-dimensional lesion mapping. J Neurosci. 2000;20(7):2683–90.

Amodio DM, Frith CD. Meeting of minds: the medial frontal cortex and social cognition. Nat Rev Neurosci. 2006;7(4):268–77.

Angelidis J, Ibrahim NA. The impact of emotional intelligence on the ethical judgment of managers. J Bus Ethics. 2011;99:111–9.

Ashar YK, Andrews-Hanna JR, Dimidjian S, Wager TD. Empathic care and distress: predictive brain markers and dissociable brain systems. Neuron. 2017;94(6):1263–73.e4.

Ashforth BE, Humphrey RH. Emotion in the workplace: a reappraisal. Hum Relat. 1995;48 (2):97–125.

Babiak P. When psychopaths go to work: a case study of an industrial psychopath. Appl Psychol Int Rev. 1995;44(2):171–88.

Babiak P, Hare R. Snakes in suits: when psychopaths go to work. New York: HarperCollins; 2006.

Bagozzi RP, Verbeke WJMI, Dietvorst RC, Belschak FD, van den Berg WE, Rietdijk WJR. Theory of mind and empathic explanations of machiavellianism: a neuroscience perspective. J Manag. 2013;39(7):1760–98.

Bandura A. Moral disengagement in the perpetration of inhumanities. Pers Soc Psychol Rev. 1999;3(3):193–209.

Barnard CI. The functions of the executives. Cambridge: Harvard University Press; 1938.

Barnett MA. Empathy and related responses in children. In: Eisenberg N, Strayer J, editors. Empathy and its development. Cambridge: Cambridge University Press; 1987.

Batson CD. These things called empathy: eight related but distinct phenomena. In: Decety J, Ickes W, editors. The social neuroscience of empathy. Cambridge, MA: MIT; 2009.

Batson DC. The empathy-altruism hypothesis: issues and implications. In: Decety J, Ickes W, editors. Empathy: from bench to bedside. Cambridge: MIT Press; 2012.

Batson CD, Klein TR, Highberger L, Shaw LL. Immorality from empathy-induced altruism: when compassion and justice conflict. J Pers Soc Psychol. 1995;68(6):1042–54.

Batson CD, Early S, Salvarani G. Perspective taking: imagining how another feels versus imaging how you would feel. Pers Soc Psychol Bull. 1997;23(7):751–8.

Batson CD, Ahmad N, Yin J, Bedell SJ, et al. Two threats to the common good: self-interested egoism and empathy-induced altruism. Pers Soc Psychol Bull. 1999;25(1):3–16.

Beetz A, Uvnäs-Moberg K, Julius H, Kotrschal K. Psychosocial and psychophysiological effects of human-animal interactions: the possible role of oxytocin. Front Psychol. 2012;3:234.

Bird FB. The muted conscience: moral silence and the practice of ethics in business. Westport: Quorum Books; 1996.

Bird G, Silani G, Brindley R, White S, Frith U, Singer T. Empathic brain responses in insula are modulated by levels of alexithymia but not autism. Brain. 2010;133(Pt 5):1515–25.

Bloom P. Against empathy: the case for rational compassion. New York: Ecco; 2016. 304 p.

Boddy CR. Corporate psychopaths, conflict, employee affective well-being and counterproductive work behaviour. J Bus Ethics. 2014;121(1):107–21.

Boyke J, Driemeyer J, Gaser C, Buchel C, May A. Training-induced brain structure changes in the elderly. J Neurosci. 2008;28(28):7031–5.

Brown TA, Sautter JA, Littvay L, Sautter AC, Bearnes B. Ethics and personality: empathy and narcissism as moderators of ethical decision making in business students. J Educ Bus. 2010;85:203–8.

Bryant BK. Mental health, temperament, family, and friends: perspectives on children's empathy and social perspective taking. In: Eisenberg N, Strayer J, editors. Empathy and its development. Cambridge: Cambridge University Press; 1987.

Bufalari I, Aprile T, Avenanti A, Di Russo F, Aglioti SM. Empathy for pain and touch in the human somatosensory cortex. Cerebral Cortex. 2007;17(11):2553–61.

Chartrand TL, Lakin JL. The antecedents and consequences of human behavioral mimicry. Annu Rev Psychol. 2013;64:285–308.

Cheng Y, Chen C, Lin CP, Chou KH, Decety J. Love hurts: an fMRI study. Neuroimage. 2010;51 (2):923–9.

Clarke J. Working with monsters. How to identify and protect yourself from workplace psychopaths. Sydney: Random House; 2005.

Cohen MA. Empathy in business ethics education. J Bus Ethics Educ. 2012;9:359–75.

Coplan A. Understanding empathy. In: Coplan A, Goldie P, editors. Empathy: philosophical and psychological perspectives. New York: Oxford University Press; 2011. p. 3–18.

Damasio AR. Emotion and human brain. Ann NY Acad Sci. 2001;935:101–6.

Damasio AR. Descartes' error: emotion, reason, and the human brain. New York: Penguin Books; 2005. 368 p.

Damasio H, Grabowski T, Frank R, Galaburda AM, Damasio AR. The return of Phineas Gage: clues about the brain from the skull of a famous patient. Science. 1994;264:1102–5.

de Vignemont F. Affective mirroring: emotional contagion or empathy? In: Nolen-Hoeksema S, Fredrickson BL, Loftus GR, Wagenaar WA, editors. Atkinson & Hilgard's introduction to psychology. 15th ed. Cengage Learning: Andover; 2009.

de Vignemont F, Singer T. The empathic brain: how, when and why? Trends Cogn Sci. 2006;10 (10):435–41.

Decety J. To what extent is the experience of empathy mediated by shared neural circuits? Emot Rev. 2010;2(3):204–7.

Decety J. The neural pathways, development and functions of empathy. Curr Opin Behav Sci. 2015;3:1–6.

Decety J, Cowell JM. Friends or foes: is empathy necessary for moral behavior? Perspect Psychol Sci. 2014a;9(5):525–37.

Decety J, Cowell JM. The complex relation between morality and empathy. Trends Cogn Sci. 2014b;18(7):337–9.

Decety J, Cowell JM. Empathy, justice, and moral behavior. AJOB Neurosci. 2015a;6(3):3–14.

Decety J, Cowell JM. The equivocal relationship between morality and empathy. In: Decety J, Wheatley T, editors. The moral brain: a multidisciplinary perspective. Cambridge: MIT; 2015b.

Decety J, Jackson PL. The functional architecture of human empathy. Behav Cogn Neurosci Rev. 2004;3(2):71–100.

Decety J, Moriguchi Y. The empathic brain and its dysfunction in psychiatric populations: implications for intervention across different clinical conditions. BioPsychoSoc Med. 2007;1:22–65.

Decety J, Yoder KJ. Empathy and motivation for justice: cognitive empathy and concern, but not emotional empathy, predict sensitivity to injustice for others. Soc Neurosci. 2016;11(1):1–14.

Decety J, Norman GJ, Berntson GG, Cacioppo JT. A neurobehavioral evolutionary perspective on the mechanisms underlying empathy. Prog Neurobiol. 2012;98(1):38–48.

Decety J, Lewis KL, Cowell JM. Specific electrophysiological components disentangle affective sharing and empathic concern in psychopathy. J Neurophysiol. 2015;114(1):493–504.

Dewey J. Human nature and conduct: an introduction to social psychology. New York: Holt; 1922.

Dietz J, Kleinlogel EP. Wage cuts and managers' empathy: how a positive emotion can contribute to positive organizational ethics in difficult times. J Bus Ethics. 2014;119(4):461–72.

Donaldson T, Preston LE. The stakeholder theory of the corporation: concepts, evidence, and implications. Acad Manag Rev. 1995;20(1):65–91.

Dutton JE, Worline MC, Frost PJ, Lilius J. Explaining compassion organizing. Adm Sci Q. 2006;51 (1):59–96.

Ebert A, Brüne M. Oxytocin and social cognition. In: Hurlemann R, Grinevich V, editors. Behavioral pharmacology of neuropeptides: oxytocin. Current topics in behavioral neurosciences, vol. 35. Cham: Springer; 2017.

Echols S, Correll J. It's more than skin deep: empathy and helping behavior across social groups. In: Decety J, editor. Empathy: from bench to bedside. Cambridge: MIT; 2012.
Eisenberg N, Eggum ND. Empathic responding: sympathy and personal distress. In: Decety J, Ickes W, editors. The social neuroscience of empathy. Cambridge: MIT; 2009.
Engen HG, Singer T. Empathy circuits. Curr Opin Neurobiol. 2013;23(2):275–82.
Eres R, Decety J, Louis WR, Molenberghs P. Individual differences in local gray matter density are associated with differences in affective and cognitive empathy. Neuroimage. 2015;117:305–10.
Fan Y, Duncan NW, de Greck M, Northoff G. Is there a core neural network in empathy? An fMRI based quantitative meta-analysis. Neurosci Biobehav Rev. 2011;35(3):903–11.
FeldmanHall O, Dalgleish T, Evans D, Mobbs D. Empathic concern drives costly altruism. NeuroImage. 2015;105:347–56.
Feshbach ND, Feshbach S. Empathy and education. In: Decety J, Ickes W, editors. The social neuroscience of empathy. Cambridge: MIT; 2009.
Gaudine A, Thorne L. Emotion and ethical decision-making in organizations. J Bus Ethics. 2001;31 (2):175–87.
Gazzaniga MS. The ethical brain. New York: Dana; 2005.
Geng Y, Zhao W, Zhou F, Ma X, Yao S, Hurlemann R, et al. Oxytocin enhancement of emotional empathy: generalization across cultures and effects on amygdala activity. Front Neurosci. 2018;12:512.
Georgi E, Petermann F, Schipper M. Are empathic abilities learnable? Implications for social neuroscientific research from psychometric assessments. Soc Neurosci. 2014;9(1):74–81.
Gleichgerrcht E, Young L. Low levels of empathic concern predict utilitarian moral judgment. PLoS One. 2013;8(4):e60418.
Goleman D. Emotional intelligence: why it can matter more than IQ. Toronto: Bantam Books; 1995.
Gonzalez-Liencres C, Shamay-Tsoory SG, Brune M. Towards a neuroscience of empathy: ontogeny, phylogeny, brain mechanisms, context and psychopathology. Neurosci Biobehav Rev. 2013;37(8):1537–48.
Hare R. Predators: the disturbing world of the psychopaths among us. Psychol Today. 1994;27 (1):54–63.
Hatfield E, Cacioppo JT, Rapson RL. Emotional contagion. Curr Dir Psychol Sci. 1993;2(3):96–9.
Hatfield E, Cacioppo JT, Rapson RL. Emotional contagion. New York: Cambridge University Press; 1994.
Hein G, Silani G, Preuschoff K, Batson CD, Singer T. Neural responses to ingroup and outgroup members' suffering predict individual differences in costly helping. Neuron. 2010;68 (1):149–60.
Hojat M, Vergare MJ, Maxwell K, Brainard G, Herrine SK, Isenberg GA, et al. The devil is in the third year: a longitudinal study of erosion of empathy in medical school. Acad Med. 2009;84 (9):1182–91.
Holt S, Marques J. Empathy in leadership: appropriate or misplaced? An empirical study on a topic that is asking for attention. J Bus Ethics. 2012;105(1):95–105.
Humphrey RH. The benefits of emotional intelligence and empathy to entrepreneurship. Entrep Res J. 2013;3:3.
Jenkins AC, Mitchell JP. Medial prefrontal cortex subserves diverse forms of self-reflection. Soc Neurosci. 2011;6(3):211–8.
Jones TM. Ethical decision making by individuals in organizations: an academy of management. Acad Manag Rev. 1991;16(2):366.
Kayworth TR, Leidner DE. Leadership effectiveness in global virtual teams. J Manag Inf Syst. 2001;18(3):7–40.
Kellett JB, Humphrey RH, Sleeth RG. Empathy and the emergence of task and relations leaders. Leadersh Q. 2006;17(2):146–62.
Kidd DC, Castano E. Reading literary fiction improves theory of mind. Science. 2013;342 (6156):377–80.

Kitcher P. The ethical project. Cambridge: Harvard University Press; 2011.

Lamm C, Meltzoff AN, Decety J. How do we empathize with someone who is not like us? A functional magnetic resonance imaging study. J Cogn Neurosci. 2010;22(2):362–76.

Lamm C, Decety J, Singer T. Meta-analytic evidence for common and distinct neural networks associated with directly experienced pain and empathy for pain. Neuroimage. 2011;54 (3):2492–502.

Law KS, Chi-Sum W, Song LJ. The construct and criterion validity of emotional intelligence and its potential utility for management studies. J Appl Psychol. 2004;89(3):483–96.

Leiberg S, Anders S. The multiple facets of empathy: a survey of theory and evidence. Prog Brain Res. 2006;156(4):419–40.

Leslie AM, Friedman O, German TP. Core mechanisms in 'theory of mind'. Trends Cogn Sci. 2004;8(12):528–33.

Lurie Y. Humanizing business through emotions: on the role of emotions in ethics. J Bus Ethics. 2004;49(1):1.

Mahsud R, Yukl G, Prussia G. Leader empathy, ethical leadership, and relations-oriented behaviors as antecedents of leader-member exchange quality. J Manag Psychol. 2010;25(6):561–77.

Mayer JD, Salovey P. What is emotional intelligence? Emotional development and emotional intelligence: educational implications. New York: Basic Books; 1997. p. 3–34.

Mayer JD, Salovey P, Caruso DR. Emotional intelligence: theory, findings, and implications. Psychol Inq. 2004;15(3):197–215.

McVea JF, Freeman RE. A names-and-faces approach to stakeholder management: how focusing on stakeholders as individuals can bring ethics and entrepreneurial strategy together. J Manag Inq. 2005;14(1):57–69.

Mencl J, May DR. The effects of proximity and empathy on ethical decision-making: an exploratory investigation. J Bus Ethics. 2009;85(2):201–26.

Mintzberg H, Simons R, Basu K. Beyond selfishness. MIT Sloan Manag Rev. 2002;44(1):67–74.

Moll J, de Oliveira-Souza R. Moral judgments, emotions and the utilitarian brain. Trends Cogn Sci. 2007;11(8):319–21.

Mora F, Segovia G, del Arco A. Aging, plasticity and environmental enrichment: structural changes and neurotransmitter dynamics in several areas of the brain. Brain Res Rev. 2007;55(1):78–88.

Muller AR, Pfarrer MD, Little LM. A theory of collective empathy in corporate philanthropy decisions. Acad Manag Rev. 2014;39(1):1–21.

Nussbaum MC. Not for profit: why democracy needs the humanities. Princeton: Princeton University Press; 2010.

Pavlovich K, Krahnke K. Empathy, connectedness and organisation. J Bus Ethics. 2012;105:131–7.

Pizarro D. Nothing more than feelings? The role of emotions in moral judgment. J Theory Soc Behav. 2000;30(4):355–75.

Preston SD, de Waal F. Empathy: its ultimate and proximate bases. Behav Brain Sci. 2002;25:1–72.

Preston SD, Bechara A, Damasio H, Grabowski TJ, Stansfield RB, Mehta S, et al. The neural substrate of cognitive empathy. Soc Neurosci. 2007;2(3–4):254–75.

Racine E. Interdisciplinary approaches for a pragmatic neuroethics. Am J Bioethics. 2008;8 (1):52–3.

Racine E. Pragmatic neuroethics. Improving treatment and understanding of the mind brain. Cambridge: MIT; 2010.

Riess H, Kelley JM, Bailey RW, Dunn EJ, Phillips M. Empathy training for resident physicians: a randomized controlled trial of a neuroscience-informed curriculum. J Gen Intern Med. 2012;27 (10):1280–6.

Ruby P, Decety J. What you believe versus what you think they believe: a neuroimaging study of conceptual perspective-taking. Eur J Neurosci. 2003;17(11):2475–80.

Salminen JK, Saarijärvi S, Äärelä E, Toikka T, Kauhanen J. Prevalence of alexithymia and its association with sociodemographic variables in the general population of Finland. J Psychosom Res. 1999;46(1):75–82.

Schlegel AA, Rudelson JJ, Tse PU. White matter structure changes as adults learn a second language. J Cogn Neurosci. 2012;24(8):1664–70.
Seo M-G, Barrett LF. Being emotional during decision making – good or bad? An empirical investigation. Acad Manag J. 2007;50(4):923–40.
Sheng F, Han S. Manipulations of cognitive strategies and intergroup relationships reduce the racial bias in empathic neural responses. NeuroImage. 2012;61(4):786–97.
Singer T, Fehr E. The neuroeconomics of mind reading and empathy. Am Econ Rev. 2005;95 (2):340–5.
Smith A. The theory of moral sentiments. London: Penguin Books; 2009 [1790].
Smith ER, Seger CR, Mackie DM. Can emotions be truly group level? Evidence regarding four conceptual criteria. J Pers Soc Psychol. 2007;93(3):431.
Smith KE, Porges EC, Norman GJ, Connelly JJ, Decety J. Oxytocin receptor gene variation predicts empathic concern and autonomic arousal while perceiving harm to others. Soc Neurosci. 2013;9 (1):1–9.
Stein SJ, Papadogiannis P, Yip JA, Sitarenios G. Emotional intelligence of leaders: a profile of top executives. Leadersh Org Dev J. 2009;30(1):87–101.
Taubert M, Draganski B, Anwander A, Muller K, Horstmann A, Villringer A, et al. Dynamic properties of human brain structure: learning-related changes in cortical areas and associated fiber connections. J Neurosci. 2010;30(35):11670–7.
Taylor GJ, Bagby MR, Parker JD. The alexithymia construct: a potential paradigm for psychosomatic medicine. Psychosomatics. 1991;32(2):153–64.
Trinkaus J, Giacalone J. The silence of the stakeholders: zero decibel level at Enron. J Bus Ethics. 2005;58:237–48.
van Baaren RB, Holland RW, Kawakami K, van Knippenberg A. Mimicry and prosocial behavior. Psychol Sci. 2004;15(1):71–4.
Ward J, Cody J, Schaal M, Hojat M. The empathy enigma: an empirical study of decline in empathy among undergraduate nursing students. J Prof Nurs. 2012;28(1):34–40.
Wilson EO. Consilience: the unity of knowledge. New York: Vintage Books; 1999.
Yoder KJ, Decety J. The neuroscience of morality and social decision-making. Psychol Crime Law. 2018;24(3):279–95.
Zahavi D, Overgaard S. Empathy without isomorphism: a phenomenological account. In: Decety J, editor. Empathy: from bench to bedside. Cambridge: MIT; 2012.
Zhong C-B. The ethical dangers of deliberative decision making. Adm Sci Q. 2011;56(1):1–25.

Neural and Behavioral Insights into Online Trust and Uncertainty

13

Uma R. Karmarkar and Adrianna C. Jenkins

Contents

Abstract

Trust is a psychological factor that can gate channels of communication, persuasion, and engagement between individuals and organizations in digital settings. Thus, understanding when and how people trust online entities has become an important issue for the technology industry. In this chapter, we provide an overview of the insights that neuroscience offers us into trust, and perhaps more importantly distrust and uncertainty, with particular attention to how these processes relate to digital spaces and cybersecurity. We use this to provide building blocks for articulating the questions that dominate the ethical responsibility of organizations in communicating with their customers in these settings.

U. R. Karmarkar (✉)
Rady School of Management, and School of Global Policy and Strategy, University of California, San Diego, San Diego, CA, USA
e-mail: ukarmarkar@ucsd.edu

A. C. Jenkins
Department of Psychology, University of Pennsylvania, Philadelphia, PA, USA

J. T. Martineau, E. Racine (eds.), *Organizational Neuroethics*, Advances in Neuroethics, https://doi.org/10.1007/978-3-030-27177-0_13

13.1 Introduction

The last few decades have seen increases in the transactions we conduct online. Whether related to business or social or personal concerns, these transactions commonly occur without face-to-face interpersonal contact. Instead, they rely on trust in e-commerce, online entities, and web platforms. Online transactions are growing rapidly in number and scope. According to recent studies, 96% of US adults with Internet access have shopped online at least once (Square and ECommerce 2017), and more than half of all travel booking from the USA now takes place online (Statistics Brain 2017). By some estimates, e-commerce is growing at a staggering rate of 23% per year (Square and ECommerce 2017). Similarly, businesses' websites increasingly link to online review repositories like Yelp and TripAdvisor, which give users access of hundreds of millions of evaluations not limited by proximity or personal connection.

The rise of online transactions has brought with it new kinds of risks and uncertainties for both businesses and their audiences. Accordingly, understanding when and how people develop trust and distrust in digital channels has become an important issue for organizations, particularly those in the technology industry. On the one hand, it is critical for firms with a digital presence to engender the trust of their customers in ways that span offline and online interactions. This is obviously even more significant for firms that only have online offerings. On the other hand, given that cybersecurity failures can often be traced to human error, firms and institutions are equally interested in how to make people selectively less trusting and more cautious. A recent investigation of cybercrimes by the FBI concluded that phishing attacks in which a criminal impersonates a company executive (Business Email Compromise attacks, or BECs) cost US businesses approximately half a billion dollars annually.

In light of these challenges, what are the factors that give rise to online trust? How might understanding these challenges inform policies that strike a balance between engendering trust to increase engagement and encouraging distrust to improve security? Converging results across multiple fields of study suggest that online trust could be conceptualized in two distinguishable ways: either as a facet of an interpersonal relationship or as an evaluation of the safety or riskiness of an online setting as a whole. The first centers on a specific target of potential trust, such as another individual or an organization. Though often arising spontaneously based on minimal information, this interagent trust can be developed through one or more interactions. This type of trust is also strongly influenced by social cues, whether from a firm itself or from others who relate to the firm. The second form centers on individuals' experience of trust more generally; it comprises a feeling of trust (or feeling of distrust) associated with online settings. This second form can depend on a variety of contextual cues, and, because of the risks and/or level of unknown factors in this relatively new way of interacting, this general form of trust may be more related to individuals' tolerance for risk and uncertainty. Both types of online trust critically rely on how people use the information they have in settings where their knowledge might be flawed or incomplete.

The importance of these issues in determining economic and welfare consequences for people engaging in online transactions raises important ethical questions about what, when, and how firms should be communicating to users. In particular, what are the trade-offs for organizations between doing what is most ethical and doing what is most effective or profitable in these contexts? What are the consequences for users if they have more information versus simply information that makes them feel more confident? Every online transaction requires some degree of trust, and the consequences of trusting or failing to trust in the wrong circumstances can be high. In these situations, companies are in a position of power relative to users, which can create a tension—whether or not it is perceived by a particular company—between helping people make better choices and taking away their agency to circumvent poorer ones. To understand the impact of such actions and the responsibilities of the firms, or other online agents, it is necessary to first understand how various types of interactions and information impact people's feelings of trust and distrust. Doing so means examining the intersection of psychology, neuroscience, economics, and ethics.

As a step toward untangling this, we discuss the wealth of research on the neurocognitive underpinnings of trust and distrust as they pertain to online transactions. Specifically, we discuss the antecedents, markers, and modulators of trust in social interactions and how they may be affected by the uncertainty inherent in online settings. This is complemented by an exploration of how uncertainty may specifically engender distrust and the importance of the amount and type of information available to a decision-maker. We apply these insights to inform questions about the ethical responsibility of online institutions in developing policies and strategies related to communicating with their customers as they seek to balance trust and mistrust in an increasingly digital world.

13.2 The Online World and Its Implications for Trust

In recent years, advances in technology and globalization have launched our species into increasingly anonymous interactions with vast numbers of individuals across great distances, rendering invalid many of the assumptions about the landscape of social and economic exchange under which the human brain evolved (Caporael 1997; Fiske 2000; Tooby and Cosmides 1990). Rather than a landscape characterized by repeated interactions with a small number of proximally located, personally known individuals, people increasingly engage in one-time interactions with a large number of geographically dispersed strangers. Perhaps nowhere are these changes more evident than in online consumer transactions.

We can illustrate some of the potential differences in trust in online versus offline transactions with the results from a recent survey related to a loss of personal information. One hundred and fifty-one individuals participated in the survey via Amazon's Mechanical Turk platform. Trust and effort to avoid repeat transactions

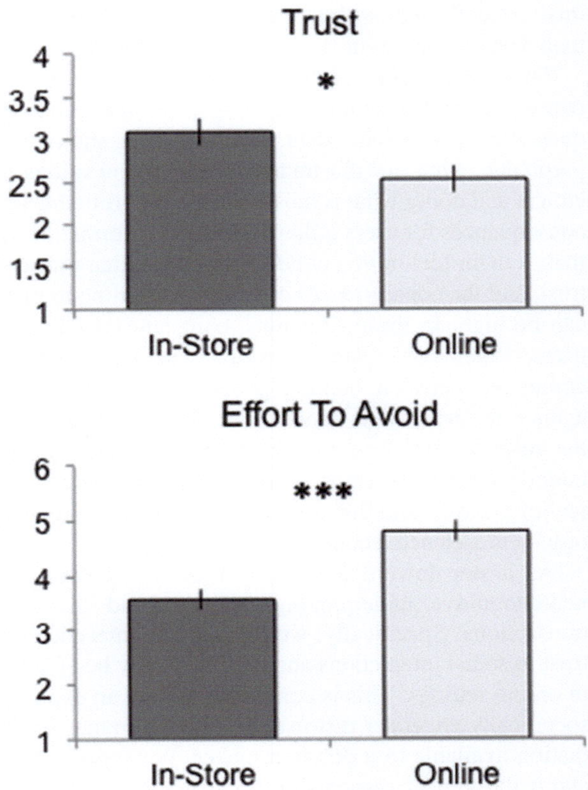

Fig. 13.1 After a security failure, survey respondents indicated they would have less trust in the store if the compromised transaction was online versus in-store ($p < 0.05$). Similarly, respondents indicated they would invest more effort in avoiding the online store compared to shopping at the brick-and-mortar location ($p < 0.001$)

were measured on 7-point Likert scales. Respondents were told that the security of their credit card information had been compromised either after an in-person transaction that occurred at a brick-and-mortar location of a particular well-known retail store (half of participants) or after an electronic transaction on that same store's website (the other half of participants). People considering the online scenario reported that they would trust the retailer significantly less after this security breach than those considering the same situation taking place in store (Fig. 13.1). Thinking about future transactions, people also indicated they were more likely to avoid shopping again at the store when it was an online purchase than they were to avoid shopping again at the same physical location. These effects indicate that it is not only how trust is formed but also how it is retained that can differ between online and offline transactions.

In the following sections, we consider three main sets of features that generally distinguish online from traditional, face-to-face transactions and the implications of what we know about the human mind and brain for trust in these settings. Specifically, we discuss the transition from proximal to distal interactions, the transition from interacting primarily with individuals (including representatives for organizations) to interacting with organizations, and the transition from interacting in familiar ways or with familiar individuals to interacting via uncertain channels with unknown entities and unfamiliar dangers. Critically, a fundamental consequence of these changes has been a general increase in the uncertainty associated with social and economic transactions.

13.2.1 From Proximal to Distal Interactions

One feature known to be important for trust that distinguishes online from traditional transactions is *psychological distance* (Liberman et al. 2007; Trope and Liberman 2010). Psychological distance refers to the degree to which another person, entity, or event seems to be removed from oneself physically (here vs. there), temporally (now vs. then), or socially (us vs. them). An abundance of research in psychology and cognitive neuroscience has demonstrated the importance of distance to a wide variety of social interactions. For example, people are more willing to help others who are physically closer to them or personally identified (Small and Loewenstein 2003), and one of the most consistent observations in behavioral science is that people behave more prosocially toward people who are socially close to them (e.g., friends, relatives, group members) than toward those who are socially distant (Cikara et al. 2011; Stotland 1969; Tajfel and Turner 1986).

In turn, recent research in cognitive neuroscience suggests various aspects of psychological distance share an underlying representation in a core network of brain regions (Addis et al. 2007; Buckner and Carroll 2007; Tamir and Mitchell 2011). More robust engagement of these regions, which tend to be engaged during social thought and decision-making, has been associated with prosocial behavior (Rameson et al. 2012; Waytz et al. 2012). Yet evidence suggests that activity in these regions diminishes with degree of social distance between oneself and others (Ames et al. 2008; Jenkins et al. 2008; Mitchell et al. 2006) and with the degree of temporal distance associated with an envisioned event (Addis et al. 2009; D'Argembeau et al. 2008; Jenkins and Hsu 2017).

Taken at face value, these findings point to the possibility that online interactions, typically characterized by both greater physical and greater social distance than their in-person counterparts, might bring with them a diminished likelihood of engaging the kinds of social cognitive processes typically associated with more proximal interactions. The consequences of this difference are not yet clear, but they represent central questions for future research. In particular, it will be important to understand to what extent people might be less trusting in online transactions on the whole versus the extent to which, in online settings, people might be less sensitive to cues that normally engender feelings of trust *or* distrust.

13.2.2 From Interacting with People to Interacting with Computers and Organizations

A second feature known to be important for trust that distinguishes online from traditional transactions is the degree to which the target of trust is, or seems to be, *human.* As more and more banking, shopping, social, and other transactions take place online, the apparent targets of trust shift from individuals to other entities, including computers and organizations as a whole. Yet these entities lack many of the characteristic features of humans—notably, biological minds and bodies—that are known to affect social cognition and decision-making (Jenkins et al. 2014; Knobe and Prinz 2008).

One of the most consistent observations in cognitive neuroscience is that thinking about other people engages a particular set of brain regions, including the medial prefrontal cortex (MPFC), temporal-parietal junction (TPJ), precuneus, and superior temporal sulcus (STS). Although the particular functions carried out by these regions remain a topic of considerable debate, their engagement during social cognitive tasks is remarkably consistent (Delgado et al. 2016; Jenkins and Mitchell 2011; Saxe 2006). In particular, these regions tend to be less engaged when people think about non-human entities, including when they think about robots and computers, than when they think about the minds of human beings—even when the same kinds of judgments are involved in both cases (Amodio and Frith 2006; Jenkins and Mitchell 2011; Saxe 2006). For example, in fMRI studies, playing a game against a human engaged these regions more than playing the same game against a computer (McCabe et al. 2001; Rilling et al. 2004), and thinking about whether a series of particular traits (e.g., “honest,” “caring”) could describe a person engaged these regions more than thinking about whether the trait could describe an inanimate object (Mitchell et al. 2002). These neural observations join an abundance of behavioral evidence that thinking, reasoning, and problem-solving about *people* may be importantly different from thinking, reasoning, and problem-solving about other sorts of entities (Jenkins and Mitchell 2011). For example, people have been shown to perform better on tests of logical reasoning when it involves reasoning about other people (Cosmides 1989), which has interesting implications for distinctions between online and offline decision-making in cases when the target of reasoning shifts from humans to non-human entities.

Despite the consistent neural and behavioral distinctions between thinking about human and non-human targets, studies have also shown that the degree to which the mind treats another entity *as though it is human* is malleable and susceptible to manipulation. This malleability works both ways. On one hand, brain regions associated with thinking about people sometimes show diminished responses to human individuals, notably including cases when those individuals are dissimilar or dehumanized (Harris and Fiske 2006; Jenkins et al. 2008; Mitchell et al. 2006). On the other hand, giving inanimate objects, human characteristics—like names, eyes, voices, and apparent feelings—can increase both people’s trust in these entities (Waytz et al. 2014) and people’s compliance with their (apparent) requests (Ahn et al. 2014). Indeed, we found evidence that people are sometimes willing to attribute mental states (e.g., thoughts, beliefs, and preferences) to a group agent as a whole, like a corporation, over and above the mental states that they attribute to the people that make up that group and that, when they do, they engage brain regions associated with thinking about individual humans (Jenkins et al. 2014).

Together, these observations raise important questions about the apparent target of trust that companies and other online entities present to users, along with its consequences for users' decisions. In particular, one main set of considerations surrounds the perils for users of failing to see the relevant target of online trust as human when they should. These would include cases in which people interact with an agent they perceive as non-human but that is driven by human decisions in reality. To the extent that an online platform fails to trigger the kinds of social cognitive processing typically associated with human interaction, users could be less attuned to the possible human motives and contingencies underlying it, potentially compromising their decision-making. In contrast, a second main set of considerations surrounds the perils for users of seeing the relevant target of online trust as human when in fact they should not. To the extent that an online platform triggers the kinds of social cognitive processing typically associated with human interaction when in fact no human decision-maker is behind the relevant aspect of a transaction, users could be more likely to assume another person could catch an error or resolve or prevent problem, potentially also compromising their decision-making. Critically, the observed malleability of social cognition, both neurally and behaviorally, suggests that organizations are in a position to nudge users' responses in both directions and thus need to be aware of the implications of their approach.

13.2.3 From Interacting with Familiar Others to Interacting with Unknown Entities

We have examined the effects of increases in the perceived distance between oneself and a target of trust and changes in the type of target of trust in digital interactions. We can expand this to also explore how people handle the elements of uncertainty that derive from interacting with unknown individuals and entities for the first time. Consumers often attempt to reduce that uncertainty or form initial judgments of trustworthiness by using outside evaluations and reviews such as Amazon's "crowdsourced" ratings of the sellers that use their retail platform. This raises the question of how much such information impacts not only the initial interaction with a new entity but also later ones as well. In a relevant study, people were asked to interact with unfamiliar individuals in a multi-round economic trust game while undergoing fMRI scanning. Participants were given "outside information" describing their partners as having a "praiseworthy, neutral, or suspect moral character" (Delgado et al. 2005). In this trust game (Berg et al. 1995; Camerer and Weigelt 1988), the participant (player 1) is given some money and asked whether they want to send part of it to their partner (player 2). The amount they send gets multiplied by some constant (e.g., tripled), and player 2 can choose how much of this they want to pay back to player 1 and how much to keep for themselves. Thus the more money player 1 sends out, the more trust they are placing in player 2 to send back funds in kind.

As might be expected, participants demonstrated larger amounts of trust (e.g., sending more money) for the good or morally praiseworthy partners initially. However, they continued to do this even after interactions showing that the morally praiseworthy partners were not more trustworthy than the neutral or morally suspect ones. Looking to the neural responses, activity in the caudate nucleus signaled responses to positive and negative feedback for partners that had been described with a neutral background. These caudate signals weren't present for partners who had been initially described as trustworthy and were only weakly present for partners with "suspect" character, suggesting that less learning, or updating, was taking place.

While research in marketing often focuses on how influential a review is for a specific transaction, this neural data suggests that the influence might persist for a long time after exposure. It is thus worth considering whether positive reviews of online entities might anchor people's trust strongly enough that they fail to update their level of trust after negative experiences, putting them at risk for similar problems in the future. This in turn raises questions about the ethical responsibilities of platforms that provide such information. If the available information about an agent's trustworthiness is inappropriately positive, it not only increases the likelihood of consumers making poor decisions but can also insulate that agent against negative repercussions of their actions. Should platforms therefore take responsibility for confirming the veracity of the reviews? If so, how should veracity be established? We can further ask what kinds of consequences should occur if a firm does take on this responsibility, but does so imperfectly. How much should they be at fault for negative outcomes due to poor management of review information?

As a recent example of the dilemmas posed above, in 2017, allegations were made that an online travel platform had been inappropriately curating information related to hotels and resorts by removing reviews detailing assaults and criminal activity that occurred at these businesses (Haag 2017). While it may be obvious that creating false reviews or curating reviews in a misleading or biased manner is normatively unethical, findings like the Delgado et al. (2005) work broaden the scope of the consequences, suggesting that more minor negative information might be ignored due to an overly positive profile on such a platform. It also suggests that significant long-term damage could be done to firms via malicious creation of negative information, which is intrinsically more difficult to overcome (Baumeister et al. 2001). In light of this, the responsibility of the information channel, or platform, again becomes an important concern to address.

13.3 Comparing Trust and Distrust

While many organizations focus their attention on online trust, one of the initial user or consumer orientations to digital transactions can be distrust. This raises the question of the degree to which trust should be considered as a single construct that ranges in valence from negative (distrust) to positive (trust) versus the degree to which trust and distrust are theoretically and functionally dissociable and/or distinct (Lewicki et al. 1998). In particular, although research on online settings suggests that trust and distrust can be separately engaged by social recommendation and graphical elements of a digital interface, respectively (Seckler et al. 2015), and though trust and distrust have also been treated as distinct elements in online banking (Benamati and Serva 2007), those approaches do not offer a broader or more generalizable insight into behavior or the underlying mechanisms.

Neural data offers an opportunity to better understand how differences between trust and distrust might reflect distinct types of processing. In one fMRI study of interindividual "snap judgment" trust, Winston et al. (2002) had participants evaluate pictures of unfamiliar people. In some trials, they were asked to rate whether the people were in high school or college, and in other trials they were asked to rate how trustworthy they perceived those individuals to be. This design ensured that participants had to attend to and engage with the face in the photo across conditions, with task differences limited to evaluatory goal. It also offered a version of online situations in which initial trust is established based on inferences drawn from minimal information and subjective interpretation.

The experimenters observed higher levels of activity in right STS for the trust trials compared to the age estimation trials, suggesting that this region was engaged specifically during the evaluation of trustworthiness, not simply during engagement with the photo in general. Notably, however, the activity in the STS region was not sensitive to the *level* of trust (i.e., higher or lower) felt toward the target (Winston et al. 2002). Thus, while STS activity tracked with thinking about trust (compared to thinking about age), it did not contain information about the trustworthiness evaluation itself, raising the possibility that increased activity in this region during trustworthiness judgments might have been driven by more general demands related to social cognition, as noted in the earlier sections of this chapter. Overall, the study did not reveal a well-defined brain area correlated with perceived trustworthiness. However, relative levels of *distrust* were significantly correlated with activity in amygdala, together with lateral OFC and middle insula. Interestingly, data from lesion patients offers a convergent anatomical account. Bilateral damage to the amygdala leads to "inhibition of distrust," such that people with such lesions perceive strangers as more trustworthy and approachable than control participants do (Adolphs et al. 1998). Similarly, patients with such damage failed to increase their distrust in a partner after betrayals in a version of the Trust Game (Koscik and Tranel 2011). Together, these finding suggest that the amygdala is important not only for the process of considering another's trustworthiness but also for the output of that consideration—i.e., the judgment of (un)trustworthiness that results from it. Perhaps most importantly, these studies point to a possible role for the amygdala in *dis*trust specifically.

These studies also offer initial evidence that distrust can indeed be represented in distinct circuitry from trust. Along such lines, Dimoka (2010) directly demonstrated a dissociation between the neural representation of trust and distrust in the context of evaluating online sellers (also showing that distrust correlated with activity in the amygdala and insula). This distinction between trust and distrust becomes a critical one when considering the organizational ethics related to online trust. It suggests the importance of considering the impact of firm actions and communications on both trust and distrust, potentially even requiring strategies specific to each. While the firm has to consider the kind of response that best fits their net goals, the ethical question related to users is whether these actions best support their overall welfare, including spillover effects with other online entities.

13.4 Neural Activity Underlying Distrust and Uncertainty

We propose that examining how people behave in response to uncertainty may be an important component in questions of both trust and distrust. Given this, let us consider what we can learn by focusing on distrust. What are the specific or even unique elements of information processing or choice that occur when a person is distrustful of a situation? We may not have evidence that a particular setting, like a retailer's e-commerce site, or a social media platform, is untrustworthy, but we have an aversion to engaging in situations where we are aware of potential negative outcomes and have incomplete information about their likelihood. Similarly, distrust may arise from unfamiliarity with a particular company (i.e., agent) or with digital forms of economic and consumer transactions (i.e., process). These kinds of situations are well described by the aversion people have to uncertainty. Informally, we can define levels of uncertainty in a particular situation as the degree to which that situation involves risk and ambiguity (Fig. 13.2). We can think of risky situations as those with known probabilities and unknown outcomes, such as the flip of a coin where there is a 50/50 likelihood of getting heads or tails. In contrast, ambiguity reflects situations with unknown probabilities and unknown outcomes, more like a game between two unknown teams where we have no idea who is more or less likely to win. Ambiguity contains elements of risk, but it may offer a better reflection of many real-world situations, since we rarely have precise statistics on how a transaction might unfold.

The importance of understanding how people experience ambiguity vis-a-vis online trust draws support from fMRI studies offering a suggestive overlap between the patterns of neural activity correlated with uncertainty and those correlated with distrust. As noted in the previous section, distrust creates increased activity in areas such as the insula and amygdala. Intriguingly, similar patterns of activity are observed in a range of studies that show amygdala activation in response to uncertainty (Herry et al. 2007; Somerville et al. 2013) and more specifically ambiguity or value in ambiguous decision-making (Hsu et al. 2005; Levy et al. 2010).

While the strength of this connection is severely limited by the bounds of reverse inference, it does at least allow us to hypothesize that there may be common mechanisms involved in ambiguity aversion and distrust. Here, reverse inference refers to inferring specific functions based on activity in a particular brain area (e.g., Poldrack 2006). Such reasoning is generally inconclusive, and can lead to fallacies without additional evidence because several different functions are often correlated with the same brain area. Along these lines, behavioral research on online shopping has shown that distrust specifically predicts reduced willingness to share information and reduced willingness to purchase. Although these have been labeled as "risky" behaviors (Mcknight and Choudhury 2006), they might be better characterized as behaviors that require engaging in an ambiguous transaction.

The parallels between distrust and aversion to ambiguity/uncertainty offer a useful framework for electronic interactions in which psychological distance can hinder interpersonal trust. In particular, they suggest that we can think of distrust of online security decisions and economic interactions as aversion to uncertainty (and ambiguity in particular). Thus by understanding how people use information to make choices under ambiguity, we may get a better understanding of how people use information to make choices in such digital settings. In addition, while trust can be considered a description of a relationship between two parties or agents, uncertainty can relate to the general perspective of any online transaction, even independently of the trustworthiness of the other party. Thus ambiguity aversion may frame or interact with trust in interactions, with a person's net response arising from some integration or competition of the two processes.

13.5 Decision-Making Under Ambiguity

One hallmark of ambiguity is the element of recognized ignorance—i.e., an awareness of one's uncertainty stems from having partial or incomplete information. A wealth of studies have shown that people strongly dislike such ignorance, which creates "ambiguity aversion" (e.g., Fox and Tversky 1995). This offers a useful model of user responses to a number of online situations. For example, in the domain of cybersecurity, people need to make decisions related to whether or not to disclose personal information and/or whether or not to offer payments in situations where they feel uncertain about the website, company, or likelihood of the desired outcome. In particular, people may also feel some ignorance about the relative likelihood that they would fall prey to malicious activity from third-party actors.

We know the absence of information can lead to net increases in decision avoidance. Beyond this, however, for any particular transaction, people can select whether or not to engage. Thus, we can ask how the information that people do have, whether offered by the firm or otherwise available, guides them in making such judgments. Do people focus on the information suggesting a more favorable or safe outcome? Or do they put more weight on information suggesting an unfavorable outcome? Answering these questions is difficult, because there is strong evidence for both possibilities. Research on optimism biases and motivated reasoning suggests that people may give more weight to favorable information (Eil and Rao 2011; Jain and Maheswaran 2000; Kunda 1990; Lord et al. 1979; Russo et al. 1996; Sharot et al. 2012; Westen et al. 2006). However, it is equally true that in financial gambles, losses can "loom larger" than gains (Kahneman and Tversky 1979) and that negative information about a consumer products or even individuals has a disproportionate influence on evaluations (Baumeister et al. 2001; Peeters and Czapinski 1990; Royzman and Rozin 2006).

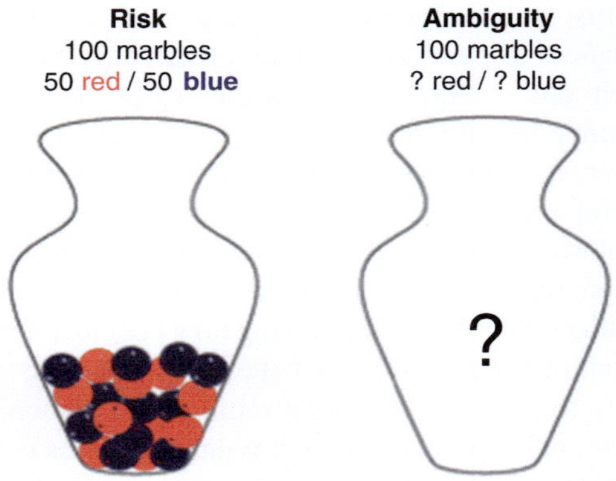

Fig. 13.2 Defining levels of uncertainty: The Ellsberg Paradox. In the classic description of the paradox of ambiguity, economist Daniel Ellsberg (1961) offered something like the following situation: Imagine you are faced with two urns. You know that both urns contain exactly 100 marbles. You know the urn on the left contains 50 red marbles and 50 blue ones. You do not know the ratio of red and blue marbles in the urn on the right, and any ratio (1:99, 40:60, 80:20, etc.) is equally possible. If you can draw a red marble from the urn (without looking) then you win a cash prize. Which do you choose from? The left urn reflects a risky decision, in which the exact probabilities are known. The right urn reflects an ambiguous decision, in which there is significant unknown information. The heart of the Ellsberg Paradox is that people overwhelmingly prefer to draw from the risky urn, even though drawing from either urn has the same probability of winning (i.e., the same expected utility). This aversion to ambiguity has been well documented across several types of experiments and decision domains (c.f., Camerer and Weber 1992; Trautmann and van de Kuilen 2015)

Our recent work has explored this question by varying the amounts of favorable and unfavorable information available in an experimental task based on the Ellsberg paradox (1969; see inset). Briefly, in the Pro/Con task (Peysakhovich and Karmarkar 2015), participants indicate how much they would be willing to pay for a ticket to play a gamble in which a poker chip is drawn from a bag of 100 red or blue chips. They are given partial but incomplete information about the color of the chips in the bag, e.g., “The bag contains at least 25 red chips and at least 35 blue chips.” Participants are told that if a red chip is drawn, they win prize money. They are also asked how likely they believe it is that a red chip will be drawn and how certain they are about their likelihood estimate.

These studies showed that people’s willingness to pay was impacted significantly more by the favorable information (e.g., number of red chips) than it was by the unfavorable information (e.g., number of blue chips). However, there was a complex set of mechanisms driving this overall behavioral effect, giving more insight into how people were using the valenced information. First, the numbers of red chips and blue chips revealed appropriately influenced estimates of the likelihood of drawing a red chip. Thus, people were not simply ignoring or discounting the unfavorable information. Second, *both favorable and unfavorable information* increased people’s subjective feelings of certainty. This is likely because both forms of information decrease ambiguity, even though they signal different outcomes. Perhaps most importantly, regardless of the estimated probability of winning, the more certain people were, the more they were willing to pay. Put another way, people placed a meaningful economic value on feeling more informed about the risks in a situation, even if that information pointed to an undesired outcome (Peysakhovich and Karmarkar 2015).

The value we place on feeling certain is a critical finding related to the types of communication that can occur between firms and users in online situations. In the Pro/Con task, there is a direct relationship between adding more information and increasing certainty in ways that mirror some forms of financial decision-making. It might then seem that all online organizations should simply offer users more information. But this ignores an important dimension of the results. It is not about whether individuals are objectively more informed. It is about their subjective feelings of certainty. In fact, complementary findings demonstrate that more information can sometimes *increase* feelings of ignorance, or reduce certainty, with meaningful behavioral consequences. For example, offering additional complicated information about investments and savings options decreases subjective knowledge and certainty, leading to less engagement with these financial opportunities (Hadar et al. 2013). In a different domain, more information about a prospective dating partner can create conflicts with expectations, reducing the interest in the person (Norton et al. 2007).

Collectively then, this research suggests that being *informative* is more useful than offering *information* from the user standpoint. In many ways, this goes against the straightforward philosophy of full transparency that can emerge from engineering and scientific organizational cultures. It additionally raises an important ethical dilemma for firms and organizations in terms of communicating with potential users. On one hand, full disclosure and access to complete information about benefits, risks, and the firm's actions might be seen as the most desirable, honest, and ethical path to take. On the other hand, an understanding of the mechanisms underlying human decision-making in such settings suggests that providing such an abundance of information is likely to be worse for a user's welfare than offering a list of key points or an otherwise curated summary and recommendations. (The latter option, of course, raises its own set of issues regarding what appropriate curation might look like.)

Terms of service (TOS) agreements offer a concrete example of this tension. Though such agreements are present to inform users of their risks and rights, the majority of users avoid exposing themselves to this avalanche of information. This can relate to finding the volume of information overwhelming and/or to wishing to avoid feelings of ignorance and uncertainty that might arise in attempting to process and comprehend all of it. Instead, most users automatically click through and accept the terms without reading all or any of the text provided. This often has direct impacts on their welfare when they later find they have agreed to relinquish rights or privacies they would not have otherwise assented to (Goel 2014). This problem is pervasive enough to have sparked community crowdsourced efforts to evaluate TOS agreements (see https://tosdr.org/). In light of this, imagine that an organization offers an accurate, transparent, and extensive TOS. Might this be considered an unethical action if it is perceived to be "too long" or "too complicated" to read? It will be important for societies, business, and governments to decide whether organizations bear the ethical responsibility to communicate information clearly and confirm users' comprehension or whether "caveat emptor" (let the buyer beware) extends to a user's responsibility to educate themselves to their satisfaction.

13.6 Conclusion

Across this chapter, we have explored several facets of trust, distrust, and uncertainty in online interactions in service of understanding the ethical questions they raise for organizations with a digital presence. In particular, we have looked to the neural and psychological mechanisms of how people build perceptions of trust and evaluate trustworthiness (and untrustworthiness) in online settings. Based on this research, we have proposed a relationship between distrust and ambiguity aversion and have thus explored the way that people make decisions in distrusted and/or uncertain environments. These research findings allow us to consider the downstream implications of the various types of actions and policies organizations might choose to pursue in online spaces.

Stepping back a bit, it is also reasonable to also consider the ethics questions raised by taking this approach in the first place. To make prescriptive recommendations, should our knowledge of human neuroscience and psychology direct organizations' ethical behavior related to online transactions? In some of the examples we have provided, the most objectively defensible approaches can clash with the reality of how people subjectively make choices and process information about others and about online situations. We can also raise practical questions about whether organizations, industries, or individuals themselves bear a responsibility to learn about how people "really" behave in these situations. Doing so is particularly challenging given that research in this field is ongoing and many of the insights discussed here are still "academic" in nature, without having been tested in noisy real-world contexts.

Overall the questions raised for organizations that engage their audience online must navigate two categories—what is "best" for users in particular situations and what is "best" for the firms. While this echoes classic debates about firms' responsibilities to shareholders versus customers, issues of privacy and cybersecurity mean that organizations cannot afford to neglect either side in their policies. Given this, the observations we have reviewed in this chapter underscore the importance of recognizing that the online environment has distinct and unique influences on human cognition and decision-making.

References

Addis DR, Wong AT, Schacter DL. Remembering the past and imagining the future: common and distinct neural substrates during event construction and elaboration. Neuropsychologia. 2007;45:1363–77.

Addis DR, Pan L, Vu MA, Laiser N, Schacter DL. Constructive episodic simulation of the future and the past: distinct subsystems of a core brain network mediate imagining and remembering. Neuropsychologia. 2009 [2014];47(11):2222–38.

Adolphs R, Tranel D, Damasio AR. The human amygdala in social judgment. Nature. 1998;393 (6684):470–4.

Ahn H-K, Kim HJ, Aggarwal P. Helping fellow beings: anthropomorphized social causes and the role of anticipatory guilt. Psychol Sci. 2014;25(1):224–9.

Ames DL, Jenkins AC, Banaji MR, Mitchell JP. Taking another person's perspective increases self-referential neural processing: short report. Psychol Sci. 2008;19(7):642–4.

Amodio DM, Frith CD. Meeting of minds: the medial frontal cortex and social cognition. Nat Rev Neurosci. 2006;7:268–77.

Baumeister RF, Bratslavsky E, Finkenauer C, Vohs KD. Bad is stronger than good. Rev Gen Psychol. 2001;5:323–70.

Benamati J "Skip", Serva MA. Trust and distrust in online banking: their role in developing countries. Inf Technol Dev 2007;13(2):161–75.

Berg J, Dickhaut J, McCabe K. Trust, reciprocity, and social history. Games Econ Behav. 1995;10 (1):122–42.

Buckner RL, Carroll DC. Self-projection and the brain. Trends Cogn Sci. 2007;11:49–57.

Camerer C, Weber M. Recent developments in modeling preferences: uncertainty and ambiguity. J Risk Uncertain. 1992;5(4):325–70.

Camerer CF, Weigelt K. Experimental tests of a sequential equilibrium reputation model. Econometrica. 1988;56(1):1–36.

Caporael LR. The evolution of truly social cognition: the core configurations model. Pers Soc Psychol Rev. 1997;1(4):276–98.

Cikara M, Bruneau EG, Saxe RR. Us and them: intergroup failures of empathy. Curr Dir Psychol Sci. 2011:149–53.

Cosmides L. The logic of social exchange: has natural selection shaped how humans reason? Studies with the Wason selection task. Cognition. 1989;31(3):187–276.

D'Argembeau A, Xue G, Lu ZL, Van der Linden M, Bechara A. Neural correlates of envisioning emotional events in the near and far future. Neuroimage. 2008;40(1):398–407.

Delgado MR, Frank R, Phelps EA. Perceptions of moral character modulate the neural systems of reward during the trust game. Nat Neurosci. 2005;8(11):1611–8.

Delgado MR, Beer JS, Fellows LK, Huettel SA, Platt ML, Quirk GJ, et al. Viewpoints: dialogues on the functional role of the ventromedial prefrontal cortex. Nat Neurosci. 2016:1545–52.

Dimoka A. What does the brain tell us about trust and distrust? Evidence from a functional neuroimaging study. Manag Inf Syst Q. 2010;34(2):11.

Eil D, Rao JM. The good news-bad news effect: asymmetric processing of objective information about yourself. Am Econ J Microecon. 2011;3(2):114–38.

Ellsberg D. Risk, ambiguity, and the savage axioms. Q J Econ. 1961;75(4):643.

Fiske ST. Stereotyping, prejudice, and discrimination at the seam between the centuries: evolution, culture, mind, and brain. Eur J Soc Psychol. 2000;30(3):299–322.

Fox CR, Tversky A. Ambiguity aversion and comparative ignorance. Q J Econ. 1995;110 (3):585–603.

Goel V. Facebook tinkers with users' emotions in news feed experiment, stirring outcry. NY Times. 2014 Jun 29.

Haag M. Tripadvisor apologizes for deleting review detailing rape at Mexican resort. NY Times. 2017 Nov 2.

Hadar L, Sood S, Fox CR. Subjective knowledge in consumer financial decisions. J Mark Res. 2013;50(3):303–16.
Harris LT, Fiske ST. Dehumanizing the lowest of the low: neuroimaging responses to extreme out-groups. Psychol Sci. 2006;17:847–53.
Herry C, Bach DR, Esposito F, Di Salle F, Perrig WJ, Scheffler K, et al. Processing of temporal unpredictability in human and animal amygdala. J Neurosci. 2007;27(22):5958–66.
Hsu M, Bhatt M, Adolphs R, Tranel D, Camerer CF. Neural systems responding to degrees of uncertainty in human decision-making. Science. 2005;310(5754):1680–3.
Jain SP, Maheswaran D. Motivated reasoning: a depth-of-processing perspective. J Consum Res. 2000;26:358–71.
Jenkins AC, Hsu M. Dissociable contributions of imagination and willpower to the malleability of human patience. Psychol Sci. 2017;28(7):894–906.
Jenkins AC, Mitchell JP. How has cognitive neuroscience contributed to social psychological theory? In: Todorov A, Fiske ST, Prentice D, editors. Social neuroscience: towards understanding the underpinnings of the social mind. Oxford: Oxford University Press; 2011.
Jenkins AC, Macrae CN, Mitchell JP. Repetition suppression of ventromedial prefrontal activity during judgments of self and others. Proc Natl Acad Sci USA. 2008;105(11):4507–12.
Jenkins AC, Dodell-Feder D, Saxe R, Knobe J. The neural bases of directed and spontaneous mental state attributions to group agents. PLoS One. 2014;9(8):e105341.
Kahneman D, Tversky A. Prospect theory: an analysis of decision under risk. Econometrica. 1979;47(2):263–92.
Knobe J, Prinz J. Intuitions about consciousness: experimental studies. Phenomenol Cogn Sci. 2008;7:67–83.
Koscik TR, Tranel D. The human amygdala is necessary for developing and expressing normal interpersonal trust. Neuropsychologia. 2011;49(4):602–11.
Kunda Z. The case for motivated reasoning. Psychol Bull. 1990;108(3):480–98.
Levy I, Snell J, Nelson AJ, Rustichini A, Glimcher PW. Neural representation of subjective value under risk and ambiguity. J Neurophysiol. 2010;103(2):1036–47.
Lewicki RJ, McAllister DJ, Bies RI. Trust and distrust: new relationships and realities. Acad Manag Rev. 1998;23(3):438–58.
Liberman N, Trope Y, Stephan E. Psychological distance. Social psychology: handbook of basic principles. 2nd ed; 2007. p. 353–81.
Lord CG, Ross L, Lepper MR. Biased assimilation and attitude polarization: the effects of prior theories on subsequently considered evidence. J Pers Soc Psychol. 1979;37(11):2098–109.
McCabe K, Houser D, Ryan L, Smith V, Trouard T. A functional imaging study of cooperation in two-person reciprocal exchange. Proc Natl Acad Sci USA. 2001;98:11832–5.
Mcknight DH, Choudhury V. Distrust and trust in B2C e-commerce: do they differ? In: Proceedings of 8th international conference on electron commerce new e-commerce innovation conqu curr barriers. Obs limitations to conduct success bus internet ACM. 2006; 482–91.
Mitchell JP, Heatherton TF, Macrae CN. Distinct neural systems subserve person and object knowledge. Proc Natl Acad Sci USA. 2002;99:15238–43.
Mitchell JP, Macrae CN, Banaji MR. Dissociable medial prefrontal contributions to judgments of similar and dissimilar others. Neuron. 2006;50:655–63.
Norton MI, Frost JH, Ariely D. Less is more: the lure of ambiguity, or why familiarity breeds contempt. J Pers Soc Psychol. 2007;92(1):97–105.
Peeters G, Czapinski J. Positive-negative asymmetry in evaluations: the distinction between affective and informational negativity effects. Eur Rev Soc Psychol. 1990;1(1):33–60.
Peysakhovich A, Karmarkar UR. Asymmetric effects of favorable and unfavorable information on decision making under ambiguity. Manage Sci. 2015;62(8):2163–78.
Poldrack RA. Can cognitive processes be inferred from neuroimaging data? Trends Cogn Sci. 2006:59–63.
Rameson LT, Morelli SA, Lieberman MD. The neural correlates of empathy: experience, automaticity, and prosocial behavior. J Cogn Neurosci. 2012;24(1):235–45.

Rilling JK, Sanfey AG, Aronson JA, Nystrom LE, Cohen JD. The neural correlates of theory of mind within interpersonal interactions. Neuroimage. 2004;22(4):1694–703.
Royzman EB, Rozin P. Limits of symhedonia: the differential role of prior emotional attachment in sympathy and sympathetic joy. Emotion. 2006;6(1):82–93.
Russo JE, Husted Medvec V, Meloy MG. The distortion of information during decisions. Organ Behav Hum Decis Process. 1996;66(1):102–10.
Saxe R. Uniquely human social cognition. Curr Opin Neurobiol. 2006:235–9.
Seckler M, Heinz S, Forde S, Tuch AN, Opwis K. Trust and distrust on the web: user experiences and website characteristics. Comput Human Behav. 2015;45:39–50.
Sharot T, Guitart-Masip M, Korn CW, Chowdhury R, Dolan RJ. How dopamine enhances an optimism bias in humans. Curr Biol. 2012;22(16):1477–81.
Small DA, Loewenstein G. Helping a victim or helping the victim: altruism and identifiability. J Risk Uncertain. 2003;26(1):5–16.
Somerville LH, Wagner DD, Wig GS, Moran JM, Whalen PJ, Kelley WM. Interactions between transient and sustained neural signals support the generation and regulation of anxious emotion. Cereb Cortex. 2013;23(1):49–60.
Square & ECommerce. Ecommerce trends: 147 stats revealing how modern customers shop in 2017. 2017. Available from https://www.bigcommerce.com/blog/ecommerce-trends/
Statistics Brain. Internet travel & hotel booking statistics. 2017. Available from https://www.statisticbrain.com/internet-travel-hotel-booking-statistics/
Stotland E. Exploratory investigations of empathy. Adv Exp Soc Psychol. 1969;4:271–314.
Tajfel H, Turner JC. The social identity theory of intergroup behavior. In: Worchel S, Austin WG, editors. Psychology of intergroup relation. Chicago: Hall; 1986. p. 7–24.
Tamir DI, Mitchell JP. The default network distinguishes construals of proximal versus distal events. J Cogn Neurosci. 2011;23(10):2945–55.
Tooby J, Cosmides L. The past explains the present. Emotional adaptations and the structure of ancestral environments. Ethol Sociobiol. 1990;11(4–5):375–424.
Trautmann ST, van de Kuilen G. Belief elicitation: a horse race among truth serums. Econ J. 2015;125(589):2116–35.
Trope Y, Liberman N. Construal-level theory of psychological distance. Psychol Rev. 2010;117 (2):440–63.
Waytz A, Zaki J, Mitchell JP. Response of dorsomedial prefrontal cortex predicts altruistic behavior. J Neurosci. 2012;32(22):7646–50.
Waytz A, Heafner J, Epley N. The mind in the machine: anthropomorphism increases trust in an autonomous vehicle. J Exp Soc Psychol. 2014;52:113–7.
Westen D, Blagov PS, Harenski K, Kilts C, Hamann S. Neural bases of motivated reasoning: an FMRI study of emotional constraints on partisan political judgment in the 2004 U.S. Presidential election. J Cogn Neurosci. 2006;18(11):1947–58.
Winston JS, Strange BA, O'Doherty J, Dolan RJ. Automatic and intentional brain responses during evaluation of trustworthiness of faces. Nat Neurosci. 2002;5(3):277–83.

Anger Expression in Organizations: Insights from Social Neuroscience 14

Esthelle Ewusi Boisvert

Anybody can become angry—that is easy, but to be angry with the right person and to the right degree and at the right time and for the right purpose, and in the right way—that is not within everybody's power and is not easy.
Aristotle, The Art of Rhetoric

Contents

E. Ewusi Boisvert (✉)
Independent Researcher, Montreal, QC, Canada

J. T. Martineau, E. Racine (eds.), *Organizational Neuroethics*, Advances in Neuroethics, https://doi.org/10.1007/978-3-030-27177-0_14

Abstract

Anger is a common emotional response to perceived provocation by another person or entity. It can emerge from feelings of frustration and manifest itself episodically or chronically. Multiple brain areas are linked to the subjective experience and expression of anger, including the amygdala-hypothalamus-periaqueductal gray system, the frontal cortices, as well as the anterior cingulate cortices. Despite its frequent negative portrayal, and general view that it can cause serious impairment, anger remains a social emotion, which can serve adaptive purposes in organizational settings. Specifically, researchers in organizational behavior have recently focused on the benefits of moral anger (i.e., when anger arises from feelings of moral violations and actions to rectify the situation ensue) in optimizing organizational functioning. The following chapter examines anger in three parts: (1) defining anger and theoretical models of anger expression; (2) the positive and negative effects of anger expression in the workplace; and (3) ways in which anger can be used constructively in organizations. The overarching aim is to examine the construct of anger in an organizational context and to present evidence from research in organizational behavior, psychology, and cognitive neuroscience that supports the idea that anger, especially moral anger, expression in the workplace can lead to better individual and collective behaviors and serve to maintain ethical organizational practices.

14.1 What Is Anger?

14.1.1 Anger: A General Definition

Anger is a common emotion, which the average adult experiences about once a day (Hendricks et al. 2013; Averill 1983). Anger can be defined as an emotional response to perceived wrongdoing by another person or entity, often characterized by, but not limited to, strong feelings of displeasure and irritation. When one has a goal in mind, and faces a challenge in achieving it, anger may arise. Anger differs from frustration in that the latter is closer to feeling upset or annoyed; frustration can be a predecessor to anger (Averill 1983). People also become angry when they feel they are being treated unfairly, when they witness injustice, or when they feel provoked (Novaco 1977). It is important to note that these are generalizations of what instigates anger in many cases, but that individuals differ in what causes anger, and how they react to or express this emotion, especially considering recommendations in an organizational context. For example, proneness to anger has been shown to vary based on individual characteristics, such as gender and socioeconomic status, with lifetime exposure to violent events being an important risk factor of anger proneness (Turner et al. 2007). Hence, because of different experiences and sociobiological factors, anger expression among individuals can vary widely in type of expression (e.g., expression, suppression), intensity (i.e., low, medium, high), and ways of expression (e.g., physical, verbal).

14.1.2 Difference Between Trait and State Anger

Within the psychology literature, anger is subdivided into trait anger and state anger. State anger is usually short-lived and presents itself along a spectrum that can range from irritation to vehement rage (Gibson and Callister 2009; Spielberger 1999). State anger can be imagined as the classic picture of a brief episode of anger on an otherwise uneventful day (Gibson and Callister 2009). Generally speaking, state or episodic anger is not a problem and can often be forgotten rather quickly, sometimes even be dismissed as trivial by the person experiencing the emotion (Elliott et al. 2015). Anger becomes problematic when it is chronic or happens multiple times a day; not only is it damaging to the angry person's health, but it contributes to damaging the said person's immediate relationships and brings toxicity to his or her work and home environments (Begley 1994; Van Kleef et al. 2004). Trait anger is characterized as a chronic tendency to experience anger, and to view situations as anger-provoking, without provocation necessarily being present (Gibson and Callister 2009; Elliott et al. 2015; Fuqua et al. 1991). The higher a person scores on trait anger, the more likely they are to report incidences of state anger as a daily occurrence (Spielberger et al. 1995). Moreover, research shows differential brain activation for individuals who score high on trait anger, with a stronger activation of the amygdala-orbitofrontal cortices (Fulwiler et al. 2012). For the purpose of this chapter, the focus will be on state anger, as being experienced by individuals in episodes, or a sequence of reactions to provocative events, rather than presenting chronically.

14.1.3 Physiology and Neural Correlates of Anger

Anger is elicited cognitively; someone will say something or do something to you that will elicit a negative response filtered through your perceptions, which will manifest in angry thoughts (Luthans 2002; Martin and Dahlen 2007). Following the formation of these ideas, or thoughts, the brain activates the autonomic nervous system, which is responsible for unconscious bodily functions, and a rush in noradrenaline prepares the body for a fight-or-flight response. Healthy participants undergoing anger induction show myriad psychobiological changes: increased heart rate, blood pressure, breathing rate, and glucose levels; tensing up of muscles; release of cortisol (stress hormone); and increased body temperature, among others (Ax 1953; Herrero et al. 2010; Hendricks et al. 2013). These psychobiological changes trigger a feedback loop; increased arousal levels can keep us angry, sometimes for extended periods of time (Mills 2005). Moderate arousal levels are necessary for optimal cognitive functioning (think about how difficult it is to work when you are too tired). On the other hand, if arousal is too intense, it can interfere with one's ability to remain concentrated and remember information clearly. When one is angry, optimal arousal levels may be exceeded, making it difficult to pursue activities, or disengage from angering situations. This also explains why it is sometimes difficult to remember details of explosive arguments, or intense anger episodes.

In the brain, anger has been associated to a differential activation of the left frontal cortex (LFC) (Harmon-Jones and Sigelman 2001; Harmon-Jones et al. 2003; Hortensius et al. 2011). Anger can also be observed as an active interaction between multiple structures, both in and outside of the limbic system. In recent years, compelling evidence has pointed to the amygdala-hypothalamus-periaqueductal gray (AHPG) system as a neural correlate of anger, as well as structures such as the FCs and the anterior and posterior cingulate cortices (Dougherty et al. 1999; Blair 2011). The AHPG system is identified as the basic threat system, one known to be critically implicated in aggression and anger. For example, as part of the limbic lobe, the dorsal anterior cingulate cortices (dACC) have been shown to be correlated with self-reported anger, or, in other words, the subjective experiencing of anger (Denson et al. 2009). The dACC are also associated with angry rumination, which means these regions remain active even after the main episode of anger has passed. In the interaction of these systems, neurotransmitters also play an important role. Serotonin is known to play a role in the regulation of aggression, sleep, and anxiety (Ursin 2002; Seo et al. 2008; Corchs et al. 2015). It is hypothesized that lower levels of serotonin might make it more difficult for the FCs and amygdala to communicate, which might make it more difficult to control or suppress anger (Blair 2011). All in all, anger uses an array of neural structures: from structures related to more reactive, instinctual behaviors (e.g., amygdala; Hendricks et al. 2013; Blair 2011) to structures associated with higher-order cognitive processes (e.g., FCs; Blair 2011).

14.2 Triggers and Expression: Cognitive Theories of Anger

14.2.1 What Triggers Anger?

Anger can emerge from internal or external sources (American Psychological Association 2017). On a daily basis, external events occur that can make us angry: someone cutting us off in traffic, running late for work, etc. External causes of anger can be the reason for episodic anger both inside and outside of work environments. Internal causes of anger are less intuitive. Anger can arise from memories of past events that made us angry, or from past trauma. Anger can also be a symptom of certain mental health disorders. The fifth and latest edition of the *Diagnostic Statistical Manual of Mental Disorders* lists "angry outbursts (with little or no provocation)" as a symptom that can present in patients who suffer from post-traumatic stress disorder or acute stress disorder (American Psychiatric Association 2013). These outbursts, or high-intensity anger episodes, are also known to sometimes be expressed by patients who suffer from major depressive disorder (Novaco 1977).

In an organizational context, internal causes of anger can often be related to perceptions of unfairness (Goldman et al. 2008); moral transgressions can even give rise to a specific subtype of anger (see subsection on moral anger in Sect. 14.3.2).

Internal causes of anger can also arise from intrinsic tendencies in irrational thinking. One example of irrational thinking patterns that can be found in the cognitive behavioral literature, and that can most certainly be applied to organizational contexts, is low frustration tolerance. Albert Ellis, psychologist and creator of Rational Emotive Behavior Therapy, argues that one of the primary reasons why it is difficult for one to change certain maladaptive behaviors might be due to low frustration tolerance (Ellis and Dryden 2007). People that engage in low frustration tolerance thinking patterns have rigid beliefs about their levels of comfort: they firmly believe that they should never engage in changes that would end up making them uncomfortable. By being motivated to avoid short-term discomfort, they remain fixed in unhealthy behavioral and cognitive patterns (Ellis and Dryden 2007). An example of this would be a supervisor responding with angry outbursts to an impediment in the workflow (because that is what they know to do and are comfortable doing) rather than making use of different, more appropriate problem-solving strategies.

14.2.2 How Do People Express Anger in the Workplace?

The subsequent section will explore two major theories in anger processing and expression: (1) Novaco's response modalities (Novaco 1977), which presents a view on the subjective experience of anger, from the perspective of the actor, or person expressing the anger; (2) Geddes and Callister's Dual Threshold Model (Geddes and Callister 2007), which presents anger expression under a social constructivist perspective, taking into account the ways in which social norms and observer perception influence anger expression in organizations. Together, both of these theories provide a brief overview of anger expression in organizations, from both an intra- and interpersonal perspective.

In order to best integrate these theoretical frameworks into an organizational context, I will present a vignette of a fictional workplace situation where anger could arise and use this vignette as an example to better contextualize both theories and hypothesize the ways in which an angry member of an organization might express her anger.

Martin and Grace: A Tale of Anger in the Workplace

> The first Monday of every month, Shirley, the manager, joins Martin and Grace, the employees, to examine work progress and establish goals for the weeks ahead. After a busy time of the year, where work responsibilities seemed to have been blurred, and much overlap occurred, this week's meeting focused on reassigning specific tasks. In the following weeks, Grace would be responsible for contacting clients and securing contracts, and Martin would be responsible for drafting the final contracts before Grace would seal the deals during her in-person meetings with the clients.
>
> In the weeks following the meeting, Grace comes across a list of potential clients she had not yet contacted and sees that their names have already been crossed off the list. When she inquires about the list to Martin, he responds that he already contacted them to try and secure contracts but obtained no contract agreement. Since Grace was the one to bring up the issue

of responsibility assignment to the last meeting, she was angry to learn that Martin did not comply with their respective duties and took over her tasks.

14.2.3 Theory 1: Anger and Novaco's Response Modalities

Raymond Novaco is an important contributor to anger management theories, which ensued from his work on cognitive therapy in the 1970s (Novaco 1975). In examining the cognitive treatment of anger and its application to depressed individuals, Novaco divided anger into three response modalities: cognitive, somatic-affective, and behavioral (Novaco 1977). All three response modalities can occur during an anger episode.

14.2.3.1 Cognitive Response Modality

According to Novaco's definition of the cognitive response modality, anger is a function of attributions that emerge from the perception of provocations (Novaco 1977). In the vignette presented above, Grace might have wrongfully attributed Martin's intentions and actions. She might have assumed that, by taking over her responsibilities, Martin was deliberately challenging Grace's professional abilities to do her work. In line with Novaco's theory, anger is known to disturb neurocognitive functions, notably by redirecting attentional resources and simplifying modes of thoughts (Lerner and Keltner 2000; Bodenhausen et al. 1994; Lerner et al. 1998). Specifically, attention toward irrelevant components of the anger-provoking situation is increased, while attention toward the important aspects of the situation is decreased (Bodenhausen et al. 1994; Lerner et al. 1998). In workplace situations, the situations are both complicated and, potentially, high stakes. This attentional redirection may have important ramifications for workplace outcomes, as important details may be missed.

What is more, when cognitively processing a situation with an added emotional lens, not only do attributions arise, but one might also have their expectations challenged. Following the meeting, Grace might have formed expectations concerning the new responsibility divide: that is, that every employee would adhere to their assigned responsibilities, at all times. By taking over Grace's responsibilities, Martin challenged that expectation. Frustration and anger are often emotions that arise when expectations are not being met. Moreover, it is in these times that our cognitions circle back to attributions, and one will attribute his or her feelings of anger to other people doing him or her wrong rather than attributing it to the internal expectations that were formed beforehand and perhaps unsuccessfully communicated (Quigley and Tedeschi 1996; Lerner and Keltner 2000).

14.2.3.2 Somatic-Effective Response Modality

Novaco describes the somatic-affective response modality of anger as the primitive, physiological expression of displeasure (e.g., racing heart), which can be exacerbated by tension and hostility (Novaco 1977; Elliott et al. 2015). When Grace first saw the list of potential clients with their names crossed off, she might have started feeling frustrated and irritated at the idea that the responsibility divide had not been respected, as discussed. Accompanying this ascending frustration, which might have turned into anger due to cognitive misattributions and challenged expectations, were physiological reactions: increases in cardiovascular activity, muscle tension, adrenaline, etc. The somatic-effective response modality serves to maintain the body's arousal levels up during an anger episode.

14.2.3.3 Behavioral Response Modality

In regard to the behavioral response modality, Novaco describes anger as manifesting through withdrawal or antagonism toward others or entities (Novaco 1977). In the fictional situation presented above, Grace's behavioral anger response could vary depending on her individual dispositions. Grace's perceptions of Martin's actions might induce a fight-or-flight response. Depending on how Grace usually displays anger, she might choose to fight the perceived origin of provocation (e.g., angry outburst) or flee the situation causing the anger (e.g., withdraw from others). Novaco explains that the reactions that can occur from the behavioral response modality can contribute to either leaving the initial instigation unchanged (in the case of withdrawal) or escalating the provocation sequence and thus exacerbating the anger (in the case of antagonism). Novaco's behavioral response modality is consistent with evidence that shows that frontal lobes activate differentially depending on the type of behavioral anger response: increased right frontal lobe activity is correlated with anger withdrawal (i.e., flight response), while increased left frontal lobe activity is associated with anger-motivated approach (i.e., fight response) (Schiff and Bassel 1996; Sobotka et al. 1992). Knowing that different behavioral responses to anger are reflected asymmetrically in the brain can be useful to creating targeted interventions for anger management (e.g., biofeedback; Achmon et al. 1989; Sobotka et al. 1992; DiGiuseppe and Tafrate 2006).

14.2.4 Theory 2: Anger and the Dual Threshold Model

The second model presented here is the Dual Threshold Model of Anger in Organizations, from Geddes and Callister (2007). Both authors' expertise on anger in the workplace adds to current theories of anger in organizations by considering the roles observers and social norms play in actor's anger expression.

14.2.4.1 Dual Threshold Model: Overview

Through the Dual Threshold Model, Geddes and Callister posit that the outcome of anger expression (positive or negative) depends on the (often implicit) established social norms regarding emotional expression in the organization (Geddes and Callister 2007). The model proposes two thresholds that can be crossed when a member of the organization expresses anger. The social norms that guide emotional expression in the organization help establish these thresholds, and the ways in which observers will react to anger expression (whether through support or sanctions), thereby influencing the type of impact anger expression can have within an organization.

14.2.4.2 Dual Threshold Model Thresholds: Expression and Impropriety

The first threshold is the expression threshold (Geddes and Callister 2007). As suggested by its title, this threshold is crossed when anger is expressed by a member of the organization to either a person or entity that is involved in provoking the anger or that can help resolve the matter. The second threshold is the impropriety threshold. This threshold is crossed when the actor's anger expression transgresses the level of anger expression that is deemed appropriate in the organization. According to Geddes and Callister, the accepted level of anger expression is established by both the actors and observers, based on the social norms that lay out the rules of emotional expression in the organization. In this sense, the Dual Threshold Model points to the importance of the actor-observer dynamic in anger expression, in addition to the actor's subjective experience of anger.

In the context of the vignette presented above, Grace could cross the first threshold by reaching out to either Martin or Shirley to express her anger at Martin's actions. The way in which Grace could cross the second threshold would depend on the social norms governing her workplace environment. We can easily imagine types of anger expression that would be deemed inappropriate in most work environments: sabotaging Martin's work, sending him violent threats, arranging a vengeful act against Martin, to name a few. Other actions—bad-mouthing Martin to Shirley, or shouting at Martin—insert themselves in a gray area of appropriateness. Without knowing which norms of emotional expression Grace's organization uses, it becomes very difficult to categorize these actions.

14.2.4.3 Dual Threshold Model: Three Types of Anger Expression

Under the model, the consequences of anger expression heavily depend on the type of expression that is used by the actor (Geddes and Callister 2007). Geddes and Callister argue that anger expression in organizations is generally more likely to engender negative outcomes than positive ones, because two out of the three possible types of anger expression lead to negative consequences:

1. Anger suppression: Occurs when the actor does not cross the first threshold. Anger suppression is a rather frequent occurrence (Booth and Mann 2005). The actor intentionally refrains from expressing anger, either by personal preference/natural inclination, because certain factors encourage it (the anger provocation comes from a superior), or due to organizational requirements (think of someone working with very young children). In the vignette presented above, Grace might decide to suppress her anger because that is the usual way she manages her anger, or because expressing her anger to Shirley or Martin would ensue severe consequences for her (e.g., suspension), as per organizational philosophy.
2. Anger expression: Occurs when the actor crosses the first threshold, but not the second. When an actor expresses anger within the acceptable norms of emotional expression, they remain in a "zone of expressive tolerance" (Geddes and Callister

2007: 723), a zone which is a safe space for positive outcomes of anger expression. This zone is flexible, and its "width" depends on established and evolving social norms of emotional expression; the larger the zone, the more space for tolerant emotion display norms (Geddes and Callister 2007; Zenteno-Hidalgo and Geddes 2012). In such a scenario, we might imagine Grace asking for a follow-up meeting with Martin and Shirley to discuss once again role assignments and where she would freely and calmly voice her feelings of anger, all while reaching out to Martin and Shirley for a proactive resolution on the situation.

3. Deviant anger expression: Occurs when the actor crosses both thresholds. The authors posit that, when an actor goes too far in their anger expression, they become the focus of the issue and make it difficult for helpful resources to focus on resolving the issue that caused the anger in the first place (Geddes and Callister 2007; Zenteno-Hidalgo and Geddes 2012). For example, in the event Grace would decide to use deviant anger expression, and cross the impropriety threshold, Shirley might decide to reprimand Grace for her actions rather than address the original situation that motivated the said actions. In this case, the conflict between Grace and Martin remains unresolved, even perhaps exacerbating Grace's already intense feelings of anger.

Geddes and Callister emphasize that, if an actor remains within the zone of expressive tolerance, then anger expression within an organization can have a positive impact on organizational well-being, with actors being more likely to express anger in less punitive ways (Geddes and Callister 2007; Geddes and Stickney 2010).

14.3 Effects of Anger in the Workplace

Until recently, organizational researchers have mostly focused on examining the negative outcomes of anger expression in the workplace; it has long been considered an emotion which needed constant regulation, and which could not contribute to a productive organizational environment (Elfenbein 2006). Despite the abundance of literature on anger's negative effects, organizational researchers have recently sought to examine anger's positive effects in organizations. Below is presented a brief overview of anger's negative and positive effects in organization, at the intrapersonal, interpersonal, and organizational levels. Considering the book's focus on organizational neuroethics, special attention is granted to moral anger and its positive effects at the organizational level.

14.3.1 Negative Effects of Anger in the Workplace

14.3.1.1 Intrapersonal Effects

As described previously, anger negatively affects physiological responses, such as blood pressure (e.g., Francis et al. 1991; Elfenbein 2006), and can also be linked to heart disease and increased feelings of hostility (Begley 1994; Lerner and Tiedens 2006). Anger can evoke an array of adjacent emotions, including fear and guilt, and can also kick-start an infinitely growing cycle of anger (Quigley and Tedeschi 1996; Geddes and Callister 2007).

Importantly, anger's physiological consequences might affect important neurocognitive functions, such as decision-making. Rapidly rising anger and increased arousal can cause steps to be skipped in the decision-making process and can influence decisions by tainting perceptions (de Castro et al. 2005). In their study of anger's effect on decision-making, Garfinkel and colleagues put it perfectly: "Emotion and cognition are dynamically coupled to bodily arousal: the induction of anger, even unconsciously, can reprioritise neural and physiological resources toward action states that bias cognitive processes" (Garfinkel et al. 2016: 150). By implicitly triggering anger in participants, the authors found participants displayed heightened blood pressure, which in turn negatively affected participant response times, a first reflection of the way in which anger affects decisions. Furthermore, results show that anger impedes on decision-making by altering the activation of the occipitotemporal and parietal gyri and significantly reducing activation in areas usually linked to decision-making (i.e., cuneus, occipital gyrus, and parietal cortex). In addition to this study by Garfinkel and colleagues, further research on anger's negative effects on decision-making and judgment (e.g., Bodenhausen et al. 1994; Garfinkel et al. 2016; Lerner and Keltner 2000) present similar results.

14.3.1.2 Interpersonal Effects

Anger can also produce negative effects at the interpersonal level (Miron-Spektor and Rafaeli 2009). One of the main social functions of anger has been identified as that of punishment (Morris and Keltner 2000). This reflects neurobiologically, as both punishment and anger share similar neural activation patterns (i.e., activation of the amygdala, medial prefrontal, and posterior cingulate cortex; Buckholtz et al. 2008). At the interpersonal level, anger has also been found to be a driving force in team conflict (Jehn 1995; Allred 1999), as well as tendencies for revenge (Tripp and Bies 1997) and blame (Aquino et al. 2001; Quigley and Tedeschi 1996). Linking back to anger's disruption of cognitive performance (Quigley and Tedeschi 1996; Garfinkel et al. 2016), anger directed at others can also perturb work performance (Van Kleef et al. 2004; Van Kleef and Côté 2007).

Anger can also play a crucial part in the relationship between emotional intelligence and expression of aggression. Anger was found to mediate the relationship between emotional intelligence and aggressive behavior (García-Sancho et al. 2016). Emotional intelligence, or the ability to regulate one's own emotions, and detect and respond to others' emotions, is an ability which is known to be an advantageous trait in organizational settings, both at the individual (Angelidis and Ibrahim 2011; García-Sancho et al. 2016; Carmeli 2003; Srivastava 2013) and team levels (Ayoko et al. 2008). Unfortunately, this means that low emotional intelligence can also engender negative interactions. At the neurobiological level, people with low emotional intelligence show a reduced activation of frontal cortices (Ayoko et al. 2008; Craig et al. 2009). The neuroscience behind emotional intelligence translates behaviorally into weaker overall social and organizational skills: people with low emotional intelligence have a lesser ability to empathize, are not as good in aligning with organizational goals, and display diminished abilities in developing social relationships (Sy and Côté 2004). Interestingly, people with low emotional intelligence will display aggressive behavior more frequently, due to their tendency to regulate their anger using angry rumination (García-Sancho et al. 2016). This might be related to problematic activation of the amygdala: evidence shows that low emotional intelligence is associated with neural lesions in the ventromedial prefrontal cortex and the amygdala, areas associated with emotional processing (Bar-On et al. 2003). In the context of a team setting where low emotional intelligence is present, it is easy to see how the high presence of angry rumination can lead to destructive behavior and attitudes (Ayoko et al. 2008; García-Sancho et al. 2016), especially knowing that emotional expression in a group setting can evolve to emotional contagion (Hatfield et al. 1994).

14.3.1.3 Organizational Effects

At the organizational level, deviant anger expression has been associated with contributing to toxic work atmospheres (Aquino et al. 2004), and even has an ability to trigger workplace violence, in extreme cases (Neuman and Baron 1998). One of the most obvious effects of anger is the emotion's potential for developing into a chronic issue (Aquino et al. 2004; Booth and Mann 2005). Furthermore, unresolved anger episodes can lead to long-term negative consequences, such as unhealthy relationships within an organization (Booth and Mann 2005). Deviant anger is also linked to reduced job satisfaction (Glomb 2002) and workplace civility (Pearson et al. 2001; Andersson and Pearson 1999), as well as resistance to organizational change (Bönigk and Steffgen 2013). Ultimately, unhealthy anger expression in the workplace can lead to irreversible consequences for an employer, such as employees quitting, or seriously considering quitting (Booth and Mann 2005).

Importantly, anger can have serious effects on leadership and perception of leadership within an organization. The effect of anger expression coming from managers is often more far-reaching than they might think: research shows that contrary to simply affecting worker productivity, anger expression coming from superiors can have long-term negative effects such as decreased trust and satisfaction from their employees, as well as reduced workplace morale (Booth and Mann 2005; Callister et al. 2017). Leaders who display anger are also unpopular among employees. Research shows that leaders who respond to an internal crisis by displaying anger are perceived less favorably by their employees than leaders who express sadness or anger and sadness simultaneously (Madera and Brent 2009). What is more, leaders' anger expression can negatively influence employee affect, leading to an increased climate of negativity in the workplace (Madera and Brent 2009).

Leaders need to be able to display emotional control and may need to suppress anger toward their employees in certain situations, as an effort to avoid confrontation, for example (Clark and Brissette 2000). Anger control in leaders can be considered of particular importance, especially when it comes to judgment and decision-making, as certain decisions carry more weight when in a leader position. Indeed, anger is known to increase one's tendency to judge others punitively, for reasons that are not always related to the original issue (Goldberg et al. 1999; Lerner et al. 1998); and so it becomes easy to see why effective leaders need not let anger get in the way of their decision-making. Emotional control (as part of high emotional intelligence) has been linked to leadership effectiveness (Goleman 1998; Gardner and Stough 2002); hence, if a leader cannot display appropriate anger, it might lead to poor management and reflect on the organizational structure as a whole.

14.3.2 Positive Effects of Anger in the Workplace

Anger is argued to have evolved to be beneficial to social environments, as it helps convey information that is important to social situations [e.g., research shows humans tend to preferentially process threatening faces (Yao et al. 2014)]. Indeed, it is important to recognize the prosocial benefits of anger in the workplace in order to encourage optimal organizational functioning through an agreeable working climate, where conflict resolution includes appropriate anger expression (Gibson and Callister 2009). Anger's positive effects in an organization at the intrapersonal, interpersonal, and organizational levels are briefly presented below.

14.3.2.1 Intrapersonal Effects

At the intrapersonal level, anger serves primarily as an internal signal that something or someone is blocking a possibility of goal achievement; anger then allows for communication about this blockage to ensue (Gibson and Callister 2009). Moreover, this identification of goal blocking allows for one to prioritize and organize their behavior in order to take the necessary steps to resolve the situation.

Anger expression's intrapersonal benefits become even more evident when contrasted with the detrimental effects of its opposite, anger suppression. Indeed, suppressing anger has been shown to promote rumination, where one constantly mentally replays the anger-provoking event without ever expressing or communicating their anger (Tice 1993). Anger suppression and rumination are not only ineffective as an emotion regulation technique (Gross 2002), but it has also been shown to be detrimental to memory and cardiovascular health (Richards and Gross 1999). Angry rumination has been shown to activate medial prefrontal cortices, as well as the hippocampus, reflecting the ways in which ruminating activates areas related to emotion regulation, encoding the provocation in memory, and problem-solving attempts (Denson et al. 2009). It is important to note that increased neural activity in these areas might reveal that cognitive resources are being spent on angry rumination rather than on using anger productively to solve the original issue, a suboptimal way, perhaps, of making use of neurocognitive resources.

14.3.2.2 Interpersonal Effects

When it comes to interpersonal benefits, "expressed anger can help interactants clarify their needs and signal boundaries of appropriate behavior" (Gibson and

Callister: 73; Denson et al. 2009). Like for many emotions, anger can incite or deter others from engaging in social behavior (Van Kleef, in Potegal et al. 2010). One of anger's main social functions would be to force another person to modify his or her behavior (Fischer and Roseman 2007). Despite the short-term aggressivity that anger expression might include, long-term effects of anger might be reconciliation and positive behavioral modification.

Anger can also be beneficial in negotiation situations, as it can be used strategically, notably to help the negotiator identify key issues for negotiation (Kopelman et al. 2006; Fischer and Roseman 2007; Van Kleef et al. 2004; Van Dijk et al. 2008). In a negotiation context, anger can become a motivator for cooperation: "Research on 'mismatching' suggests that negotiators become more cooperative to the extent that they perceive their partner as tough and intransigent, because they are motivated to avoid an unprofitable impasse, [thus making] negotiators respond in a conciliatory manner to their opponent's anger expressions" (Van Kleef, in Potegal et al. 2010: 551). Sinaceur and Tiedens have also shown that anger expression can lead to positive outcomes, as it "increases [the] expressers' ability to claim value in negotiations, when [. . .] the recipients of these expressions believe they have poor alternatives" (Sinaceur and Tiedens 2006: 314).

14.3.2.3 Organizational Effects

At the organizational level, anger is considered one of the main motivators in addressing issues of unfairness (Bies 1987). This is especially relevant, considering that perceived injustice is one of the most salient triggers of anger (e.g., Barclay et al. 2005). Researchers in organizational behavior have emphasized that anger in the workplace can be beneficial when used within its intended role as an emotion of moral justice that provokes us to confront an obstacle or offender to change the behavior of another (Morris and Keltner 2000; Barclay et al. 2005). Taking this into account, as well as the overarching theme of the book, I have decided to dedicate this section of anger's positive organizational effects on a subtype of anger called moral anger. As the prosocial effects of moral anger have been a recent interest in organizational behavior and organizational ethics research, I believe they deserve special attention in this chapter. Indeed, researchers in organizational behavior and organizational ethics such as Lindebaum and Geddes (2016) argue that allowing morally motivated anger to be expressed in the workplace can lead to better individual and collective behaviors and serve to maintain ethical organizational practices. Below is a brief introduction to moral emotions and moral anger, as well as examples of the way in which moral anger can motivate ethical behavior.

14.3.2.4 What Is a Moral Emotion? What Is Moral Anger?

Jonathan Haidt, a social psychologist specialized in the psychology of morality, defines moral emotions as "the emotions that respond to moral violations, or that motivate moral behavior" (Haidt 2003: 853). He specifies that anger is "the most underappreciated moral emotion" (p. 856). Indeed, research in personality and social psychology shows that anger arises not only from frustration but oftentimes from feelings of betrayal, immorality, and unfairness (Shaver et al. 1987; Baumeister et al.

1990; Scherer 1997). According to Haidt, it is important to move beyond our traditional characterization of angry violence and realize how anger can manifest in cases of moral justice: when standing up for others in the face of injustice, or demanding justice for ourselves (Haidt 2003; Rozin et al. 1999). Generally, as long as the goal of the expressed emotion is to benefit others, then this emotion can be deemed moral in nature and, therefore, can be useful and positive in the workplace environment.

In contrast to personal anger, which emerges from the sense that individual goals are being obstructed, moral anger is not self-serving but rather responds to unfairness toward others. Within an organizational context, anger has been deemed capable of restoring social order (Lindebaum and Geddes 2016), in contrast to the usual view that anger can cause social disorder (Holt and den Hond 2013). An essential component of moral anger is that the feeling and expression of the emotion are strongly linked to wanting to benefit others, more than oneself. According to Lindebaum and Geddes, the concept of moral anger encompasses three crucial components: (1) a want for one's anger and accompanying actions to benefit others; (2) positive intentions throughout the expression of anger (i.e., no intent of carrying harm or aggression toward others); and (3) a behavioral manifestation of the anger that would, ideally, lead to positive outcomes (Lindebaum and Geddes 2016). Lindebaum and Geddes emphasize that, based on Rest's concept of moral behavior (Rest 1986), "if an individual experiences anger as a result of moral value violations, but does not express and act upon this emotion, then we cannot speak of moral anger. Moral anger cannot remain as a mere cognitive feeling; it necessarily prompts some form of expression and action" (Lindebaum and Geddes 2016: 742).

As moral anger is a rather recently developed concept, little is known about its neural correlates, especially when it comes to organizational contexts. Most neuroscience research into moral emotions investigates other emotions (e.g., shame, guilt, disgust), and some researchers have found that anger and disgust activate similar brain areas in response to immoral situations (Moll et al. 2005), although other evidence shows differential activation (Vytal and Hamann 2010).

When looking into research on the neuroscience of moral emotions and cognition, we find that authors in the field often speak of indignation/anger (i.e., a moral emotion characterized as other critical; a sentiment which arises when a third party commits an intentional moral transgression; Moll and Schulkin 2009). Based on its definition, the concept of indignation/anger seems to be the closest comparison that can be used, if one is to elaborate on the neural correlates of moral anger. In their investigation of the cognitive neuroscience of moral emotions, Moll and colleagues found that indignation activates the lateral divisions of the orbitofrontal cortices, the anterior PFC, the anterior insula, as well as the anterior cingulate cortex (for additional evidence, see Blair et al. 1999; Moll et al. 2003). Interestingly, in addition to the areas that were identified as neural correlates of anger (i.e., frontal and anterior cingulate cortices), it seems that the moral dimension of indignation also triggers activation in the anterior insula, a region associated with emotional awareness, decision-making, and even perhaps personal moral judgment (Moll et al. 2003; Gu et al. 2013; Greene et al. 2004). The different investigations by Moll and others do not seem to detect strong activation in deeper regions of the brain (e.g., amygdala), perhaps pointing to indignation as stemming from higher-order processes coming primarily from orbitofrontal areas. Future research is needed to investigate the ways in which moral anger's neural activations differ from personal anger and other moral emotions.

14.3.2.5 Organizational Effects of Moral Anger: Can Anger Motivate Ethical Behavior?

Organizational ethics can be defined as a set of principles and guidelines that influence a business and its employees to act ethically (Valentine 2014). Overall, organizational ethics play an important role in maintaining a company's integrity and optimal functioning (Sims 1991; Koh and El'Fred 2001). Can an organization be considered to show integrity and ethics if its practices are not considered so? As Geddes and Lindebaum point out, [. . .] in many organizations, [the] motivation to act upon one's moral anger is regularly and actively discouraged or even suppressed for fear of social punishment" (Geddes and Lindebaum 2014: 5; Greene et al. 2004) Such practices as actively discouraging the expression of negative emotions not only are detrimental to employees directly but also impede on organizational transparency and overall functioning.

Moral anger can be a catalyst for ethical behavior and can prevent organizational wrongdoings (Jagannathan and Rai 2016). It can motivate employees to go beyond monitoring an organization's adherence to ethical and moral practices and encourage employees to act in the face of injustice (even at personal risk) and protect the people that could be at a disadvantage due to a failure of the organization to maintain ethical practices (Jagannathan and Rai 2016; Lindebaum and Geddes 2016; Lindebaum et al. 2016). Lindebaum, Geddes, and Gabriel emphasize that moral anger has "both informational (i.e. involving an attribution of blame) and energic (i.e. motivational) value, which jointly determine if, how, and by whom restorative action may be taken, while providing the impetus and motivation for undertaking such action" (Lindebaum et al. 2016: 5). What is more, moral anger, in contrast to deviant, or hostile anger, is "[. . .] fundamentally dialogical—seeking to establish coalitions of dissent to violence." (Lindebaum et al. 2016: 7). A prime example of the way in which moral anger can motivate ethical behavior, and ultimately encourage integrity in organizational ethics, is the denouncing of inappropriate sexual conduct in the workplace.

14.3.2.6 Moral Anger: The Cases Against Harvey Weinstein

As recently as October 2017, a report in the *New York Times* detailed the stories of actresses Rose McGowan and Ashley Judd, among others, who allege that Harvey Weinstein, Hollywood film mogul, sexually harassed and/or assaulted them (British Broadcasting Corporation 2017; Kantor and Twohey 2017). For weeks following the publication of the story, women were coming forward to describe allegations of sexual misconduct, harassment, and assault (British Broadcasting Corporation 2017). A report by *The New Yorker* explains that Weinstein made use of questionable practices to cover up sexual abuse claims, notably through the help of private investigators.

The stories have been reported by dozens of alleged victims so far and span nearly three decades (British Broadcasting Corporation 2017). To date, Weinstein has been charged with multiple counts of sexual abuse on two women. The case of Weinstein has precipitated a national reckoning over sexual misconduct in the workplace in the United States. What these women did was use their moral anger to make sure that the way they were sexualized and victimized—as vulnerable and young actresses, powerless in the face of institutional and gender dynamics—would not happen again, or to other people in similar vulnerable situations. Their hope to reestablish a sense of organizational ethics in Hollywood studios is what fuelled their anger. They were successful in encouraging a host of men and women to come forward and denounce similar situations in which they were victimized (Khomami 2017).

14.4 How to Foster Positive and Moral Anger in Organizations: General Recommendations

Geddes, Callister, and Gibson suggest that "[. . .] the opportunity to express anger when it's deemed necessary can and should be part of ongoing efforts to improve employee well-being and personal fulfillment" (Geddes et al. 2018: 26). Anger can be a complex emotion, sometimes difficult to decode even to the person experiencing it, but understanding how to best control, interpret, and communicate it can lead to anger being used constructively (Elliott et al. 2015). Below is an overview of some of the elements that can contribute to using anger to promote positive outcomes.

14.4.1 Recognizing Anger Internally and Externally

One of the first steps in choosing to use anger constructively, rather than destructively, is to understand the impact we want our anger to have and how we want ourselves and others to feel after our angry episode is over (Elliott et al. 2015). Essentially, it is crucial for employees and supervisors to understand that it is okay to feel angry at work, and even about work, but it is equally important to know how to use anger prosocially, in conflict resolution settings. In doing so, one must recognize the possible outcomes of using anger positively, for example, feeling better about the people involved in the conflict, developing the ability to prevent future conflicts, and adjusting one's biased perceptions of the situation.

Once the organizational team has recognized the positive impacts that anger can have in conflict resolution, it is important to develop anger acknowledgment skills. As mentioned beforehand, people experience anger differently. This variation in expression can make it difficult for one to consistently recognize when anger is being expressed. Moreover, additional variables such as culture can complicate the detection of emotions in others (a point which is relevant considering the increasing diversity in workplaces). Research in cultural psychology shows that participants are less accurate in identifying emotions in members from different cultures (Elfenbein and Ambady 2003). It remains that leaders who are more communal, who attend to individuals and their experiences, and who can appreciate individual feelings are more likely to promote healthy interpersonal relationships in the workplace (Ayoko et al. 2014). It is even possible that one has difficulty recognizing anger in themselves, especially if it remains an internal rumination (Kassinove 2017). Through encouraging the expression of anger, it is equally important to encourage all parties to learn about how and when they experience different types of anger and how to best express it. Organizations may want to develop internal training workshops where they would outline the difference between prosocial and deviant anger and give examples of varying situations where anger expression is acceptable. Workshops like these might be especially useful if an organization condones moral anger expression, as it can lay guidelines for employees to recognize and express moral anger appropriately.

14.4.2 Need for Space, Communication, and Cooperation

14.4.2.1 The Dual Threshold Model and the Zone of Expressive Tolerance

One of the main aspects surrounding anger expression in organizations is the norms of emotional expression. Indeed, for anger expression to be deemed appropriate, it needs to remain within the norms of appropriateness outlined in the organization (Gibson and Callister 2009; Ayoko et al. 2014). Referring back to the Dual Threshold Model, it is important to remember that organizational norms play an important role in the culture of tolerance surrounding anger expression (Geddes and Callister 2007). In order for anger to be used constructively, organizations need to allow appropriate "space" for anger expression to occur: the larger the space, the more tolerant the organization is of anger expression (Geddes and Callister 2007; Callister et al. 2017; Geddes et al. 2018). A larger "zone of expressive tolerance" leaves more space for dyads or groups to cooperate toward conflict resolution and can even foster a more supportive (rather than sanctioning) environment for dealing with deviant anger expression (Callister et al. 2017). An example of measures that demonstrate a large zone of expressive tolerance might be whistleblowers protection programs, or anonymous denunciation services within organizations. These types of initiatives or policies encourage employees to voice out issues and can encourage, among other things, the expression of moral anger.

Of course, there are risks associated with too large a space, such as anger expression provoking angry responses (Callister et al. 2017). Also, there are factors in organizations that make it so that the space for anger expression is not necessarily the same for every member of the organization. For example, research shows that anger expression is better received when it comes from men than women (Brescoll and Uhlmann 2008; Brescoll 2016; Ragins and Winkel 2011) or that superiors oftentimes have a larger space for expression than their subordinates (Gibson and

Callister 2009; Callister et al. 2017). Hence, it is important for organizations to foster an atmosphere that is open-minded and inclusive. This type of atmosphere would likely comport norms that would be conducive to creating a large zone of expressive tolerance, thereby resulting in an environment where anger expression is positive.

14.4.2.2 Cooperation and Communication

Cooperation between parties involved in conflicts is a crucial point in managing anger appropriately (Kassinove 2017; Elliott et al. 2015). When one expresses anger to someone relevant to the situation (i.e., the person concerned with the issue, or a resource that can help resolve the conflict), improvements in conflict occur (Geddes and Stickney 2010; Stickney and Geddes 2014). Moreover, research shows that healthy superior-subordinate relationships are strongly linked to prosocial upward dissent and better conflict resolution (Redmond et al. 2016). Communication and cooperation can help diffuse anger by contributing to shifting the focus from the person to the problem and, ultimately, enhance problem-solving efficiency. Hence, it becomes important for management teams to always ask: How can we make the situation better, together? How can we make sure all parties are involved in the conversation?

In making sure anger is communicated in the most optimal way possible, it is important to express anger verbally (rather than physically) and keep the anger intensity at low levels (Gibson and Callister 2009). Furthermore, it is important to be unafraid to pause discussions and resume them at a later time; immediately dealing with an anger episode might not always be the best solution (Kassinove 2017; Elliott et al. 2015). Avoiding interactions and taking time to reflect and formulate our thoughts for discussions allow for anger to dissipate, and for judgment and decision-making to become clearer, thus leading to proactive problem-solving (Elliott et al. 2015). This is one benefit of weekly meetings, where past situations can be brought up and examined after the heat of the moment has passed. What is more, bringing anger levels down to frustration even demonstrates reduced activation levels in the neural areas related to anger (Pawliczek et al. 2013).

Lastly, it is important for management teams to acknowledge that themselves and their employees have made progress both during and after conflict resolution. Positive reinforcement of adaptive organizational behaviors in the workplace has been shown to contribute to overall better organizational environments and performance enhancement; encouraging and applauding the positive use of anger in the workplace can be one of the ways to do so (Luthans 2002).

14.4.3 Regulating Anger Expression: Emotion Control and Modifying Internal Cognitions

14.4.3.1 Emotional Control

Regulating deviant anger expression through anger control has been shown to yield positive outcomes of anger expression (Gibson and Callister 2009). As seen previously, emotional regulation or emotional control is related to high emotional intelligence, an ability that is beneficial to optimal functioning within an organization (Ayoko et al. 2008). In contrast to anger suppression, where one will intentionally hold in feelings of anger, anger control involves monitoring and preventing excessive or deviant anger expression (Spielberger et al. 1995; Ayoko et al. 2008). Controlling deviant anger is known to be key in conflict resolution (Druskat 1999; Cherniss and Goleman 2003) and has been linked to individual and group performance (Gross 2002; Kelly and Barsade 2001), positive managerial outcomes (Ayoko et al. 2014), and overall organizational harmony (Weinstein et al. 1988).

At the intrapersonal level, it then becomes important for individuals to use "strategic emotional intelligence" (Troth, Jordan and Westerlaken in Ayoko et al. 2014: 257) to make a conscious effort to manage emotions in the workplace (Mayer et al. 2008). Through training in the workplace, it is possible to develop higher emotional intelligence (Cherniss and Goleman 2003; Cherniss and Adler 2000), with long-term, sustainable changes in emotional intelligence being an attainable goal for most individuals (Cherniss and Goleman 2003). Most interventions include components on self-awareness, emotional self-regulation, motivation, and general social skills. In their model for promoting emotional intelligence, Cherniss and Goleman emphasize the importance of setting training in a positive environment (i.e., a safe, social environment, with strong leadership), as well as gauging readiness, encouraging autonomy, setting specific goals, and providing feedback throughout the training (Cherniss and Goleman 2003).

14.4.3.2 Cognitive Control

In his advice about how to deal with anger, Howard Kassinove, director of Hofstra University's Institute for the Study and Treatment of Anger and Aggression, explains that a key tool is cognitive restructuring (Kassinove 2017). As explained through Novaco's cognitive response modality, anger experiences are often associated to misattributions and challenged expectations, which eventually form into cognitive distortions that contribute to exacerbating the feelings of anger. Cognitive restructuring is the practice of reanalyzing aversive situations in order to perceive them more realistically. In such a situation, role reversal can be a powerful tool. Thinking back to the fictional vignette, Grace's anger might have subsided if she changed her perceptions of Martin's actions. If Grace thought "Maybe he just wanted to help because he noticed I was overwhelmed" rather than "He only did this because he thinks I am incompetent," it might have contributed to reducing feelings of anger.

In better managing anger episodes, Kassinove also puts emphasis on learning and accepting that one can cope with negative situations, rather than believing that certain situations are unsurmountable (Kassinove 2017). This idea ties back to low frustration tolerance thinking patterns, where one will believe that negative situations are unsolvable. Of course, one should not deny that negative situations are bad, but it might also be helpful to view such situations as opportunities to develop coping and problem-solving skills.

14.5 Conclusion

The overarching aim of this chapter was to examine the construct of anger within the organizational behavior, psychology, and cognitive neuroscience literature and present its implications in an organization, with a focus on its positive outcomes. For the purpose of this chapter, anger was defined as the subjective experience of anger arising from being provoked by a third party or going through goal obstruction. Triggers of anger, both within and outside of organizations, can range in causes, from internal events, such as memories, to external events, such as being reprimanded by a superior. Importantly, anger can disturb important neurocognitive functions that are crucial to optimal organizational functioning, such as attention, judgment, and decision-making. Managers and employees can learn to use anger constructively, mainly through fostering a stronger sense of emotional intelligence. A growing literature in organizational behavior supports the idea that anger (and especially moral anger) can yield beneficial consequences for organizational transparency and organizational ethics. The lack of literature on the neuroscience of moral anger suggests that future research should examine the neural correlates of this subtype of anger, which could enhance understanding of the construct and, ultimately, better support current theories of organizational behavior and organizational ethics.

Acknowledgments This chapter benefited significantly from the insights and comments from Tovah Cowan, B.Sc.; Julia Pinheiro Carvalho, B.Sc.; and Lydia Risi, B.A., as well as editors Drs. Joé T. Martineau and Eric Racine. I thank them for their tremendous help.

References

Achmon J, Granek M, Golomb M, Hart J. Behavioral treatment of essential hypertension: a comparison between cognitive therapy and biofeedback of heart rate. Psychosom Med. 1989;51(2):152–64.

Allred KG. Anger and retaliation: toward an understanding of impassioned conflict in organizations. Res Negot Organ. 1999;7:27–58.

American Psychiatric Association. Diagnostic and statistical manual of mental disorders. 5th ed. 2013.

American Psychological Association. Controlling anger before it controls you [Internet]. American Psychological Association. (cited 2017 Dec). Available from: http://www.apa.org/topics/anger/control.aspx

Andersson LM, Pearson CM. Tit for tat? The spiraling effect of incivility in the workplace. Acad Manag Rev. 1999;24(3):452.

Angelidis J, Ibrahim NA. The impact of emotional intelligence on the ethical judgment of managers. J Bus Ethics. 2011;99(S1):111–9.

Aquino K, Tripp TM, Bies RJ. How employees respond to personal offense: the effects of blame attribution, victim status, and offender status on revenge and reconciliation in the workplace. J Appl Psychol. 2001;86(1):52–9.

Aquino K, Douglas S, Martinko MJ. Overt anger in response to victimization: attributional style and organizational norms as moderators. J Occup Health Psychol. 2004;9(2):152–64.

Averill JR. Studies on anger and aggression. Implications for theories of emotion. Am Psychol. 1983;38(11):1145–60.

Ax AF. The physiological differentiation between fear and anger in humans. Psychosom Med. 1953;15(5):433–42.

Ayoko OB, Callan VJ, Härtel CEJ. The influence of team emotional intelligence climate on conflict and team members' reactions to conflict. Small Group Res. 2008;39(2):121–49.

Ayoko OB, Ashkanasy NM, Jehn KA. Handbook of conflict management research. Northampton: Edward Elgar; 2014.

Barclay LJ, Skarlicki DP, Pugh SD. Exploring the role of emotions in injustice perceptions and retaliation. J Appl Psychol. 2005;90(4):629–43.

Bar-On R, Tranel D, Denburg NL, Bechara A. Exploring the neurological substrate of emotional and social intelligence. Brain. 2003;126(Pt 8):1790–800.

Baumeister RF, Stillwell A, Wotman SR. Victim and perpetrator accounts of interpersonal conflict: autobiographical narratives about anger. J Pers Soc Psychol. 1990;59(5):994–1005.

Begley TM. Expressed and suppressed anger as predictors of health complaints. J Organ Behav. 1994;15(6):503–16.

Bies RJ. The predicament of injustice: the management of moral outrage. In: Cummings LL, Staw BM, editors. Research in organizational behavior. Greenwich: JAI Press; 1987. p. 289–319.

Blair RJR. Considering anger from a cognitive neuroscience perspective. Wiley Interdiscip Rev Cogn Sci. 2011;3(1):65–74.

Blair RJR, Morris JS, Frith CD, Perrett DI, Dolan RJ. Dissociable neural responses to facial expressions of sadness and anger. Brain. 1999;122(5):883–93.

Bodenhausen GV, Sheppard LA, Kramer GP. Negative affect and social judgment: the differential impact of anger and sadness. Eur J Soc Psychol. 1994;24(1):45–62.

Bönigk M, Steffgen G. Effects of habitual anger on employees' behavior during organizational change. Int J Environ Res Public Health. 2013;10(12):6215–34.

Booth J, Mann S. The experience of workplace anger. Leadersh Org Dev J. 2005;26(4):250–62.

Brescoll VL. Leading with their hearts? How gender stereotypes of emotion lead to biased evaluations of female leaders. Leadersh Q. 2016;27(3):415–28.

Brescoll VL, Uhlmann EL. Can an angry woman get ahead? Status conferral, gender, and expression of emotion in the workplace. Psychol Sci. 2008;19(3):268–75.

British Broadcasting Corporation. Harvey Weinstein timeline: How the scandal unfolded [Internet]. BBC News. 2017 (cited 2017 Dec). Available from http://www.bbc.com/news/entertainment-arts-41594672

Buckholtz JW, Asplund CL, Dux PE, Zald DH, Gore JC, Jones OD, et al. The neural correlates of third-party punishment. Neuron. 2008;60(5):930–40.

Callister RR, Geddes D, Gibson DF. When is anger helpful or hurtful? Status and role impact on anger expression and outcomes. Negot Confl Manag Res. 2017;10(2):69–87.

Carmeli A. The relationship between emotional intelligence and work attitudes, behavior and outcomes. J Manag Psychol. 2003;18(8):788–813.

Cherniss C, Adler M. Promoting emotional intelligence in organizations: make training in emotional intelligence effective. American Society for Training and Development; 2000.

Cherniss C, Goleman D. The emotionally intelligent workplace: how to select for, measure, and improve emotional intelligence in individuals, groups, and organizations. New York: Wiley; 2003.

Clark MS, Brissette I. Relationship beliefs and emotion: reciprocal effects. In: Frijda NH, Manstead ASR, Bern S, editors. Emotions and beliefs. Cambridge: Cambridge University Press; 2000. p. 212–40.

Corchs F, Nutt DJ, Hince DA, Davies SJC, Bernik M, Hood SD. Evidence for serotonin function as a neurochemical difference between fear and anxiety disorders in humans? J Psychopharmacol. 2015;29(10):1061–9.

Craig A, Tran Y, Hermens G, Williams LM, Kemp A, Morris C, et al. Psychological and neural correlates of emotional intelligence in a large sample of adult males and females. Pers Individ Dif. 2009;46(2):111–5.

de Castro BO, Merk W, Koops W, Veerman JW, Bosch JD. Emotions in social information processing and their relations with reactive and proactive aggression in referred aggressive boys. J Clin Child Adolesc Psychol. 2005;34(1):105–16.

Denson TF, Pedersen WC, Ronquillo J, Nandy AS. The angry brain: neural correlates of anger, angry rumination, and aggressive personality. J Cogn Neurosci. 2009;21(4):734–44.

DiGiuseppe R, Tafrate RC. Anger treatment for adults: a meta-analytic review. Clin Psychol: Sci Pract. 2006;10(1):70–84.

Dougherty DD, Shin LM, Alpert NM, Pitman RK, Orr SP, Lasko M, et al. Anger in healthy men: a pet study using script-driven imagery. Soc Biol Psychiatry. 1999;46:466–72.

Druskat VUWSB. The link between emotions and team effectiveness: how teams engage members and build effective task processes. Acad Manag Proc. 1999;1999(1):L1–6.

Elfenbein HA. Emotion in organizations: a review and theoretical integration in stages. SSRN Electron J [Internet]. 2006. Available from https://doi.org/10.2139/ssrn.942383

Elfenbein HA, Ambady N. Universals and cultural differences in recognizing emotions. Curr Dir Psychol Sci. 2003;12(5):159–64.

Elliott CH, Smith LL, Doyle Gentry W. Anger management for dummies. New York: Wiley; 2015.

Ellis A, Dryden W. The practice of rational emotive behavior therapy. 2nd ed. New York: Springer; 2007.

Fischer AH, Roseman IJ. Beat them or ban them: the characteristics and social functions of anger and contempt. J Pers Soc Psychol. 2007;93(1):103–15.

Francis RA, Ernst FA, Nevels H, Lemeh CA. The relationship of blood pressure to a brief measure of anger during routine health screening. J Natl Med Assoc. 1991;83(7):601–4.

Fulwiler CE, King JA, Zhang N. Amygdala-orbitofrontal resting-state functional connectivity is associated with trait anger. Neuroreport. 2012;23(10):606–10.

Fuqua DR, Leonard E, Masters MA, Smith RJ, Campbell JL, Fischer PC. A structural analysis of the state-trait anger expression inventory. Educ Psychol Meas. 1991;51(2):439–46.

García-Sancho E, Salguero JM, Fernández-Berrocal P. Angry rumination as a mediator of the relationship between ability emotional intelligence and various types of aggression. Pers Individ Dif. 2016;89:143–7.

Gardner L, Stough C. Examining the relationship between leadership and emotional intelligence in senior level managers. Leadersh Org Dev J. 2002;23(2):68–78.

Garfinkel SN, Zorab E, Navaratnam N, Engels M, Mallorquí-Bagué N, Minati L, et al. Anger in brain and body: the neural and physiological perturbation of decision-making by emotion. Soc Cogn Affect Neurosci. 2016;11(1):150–8.

Geddes D, Callister RR. Crossing the line(s): a dual threshold model of anger in organizations. Acad Manage Rev. 2007;32(3):721–46.

Geddes D, Lindebaum D. In defense of anger: the significance of an under-appreciated moral emotion. AMPROC. 2014;2014(1):10390.

Geddes D, Stickney LT. The trouble with sanctions: organizational responses to deviant anger displays at work. Hum Relat. 2010;64(2):201–30.

Geddes D, Callister R, Gibson D. A message in the madness: functions of workplace anger in organizational life. Acad Manag Perspect. 2018;amp.2016.0158.

Gibson DE, Callister RR. Anger in organizations: review and integration. J Manage. 2009;36 (1):66–93.

Glomb TM. Workplace anger and aggression: informing conceptual models with data from specific encounters. J Occup Health Psychol. 2002;7(1):20–36.

Goldberg JH, Lerner JS, Tetlock PE. Rage and reason: the psychology of the intuitive prosecutor. Eur J Soc Psychol. 1999;29(56):781–95.

Goldman BM, Slaughter JE, Schmit MJ, Wiley JW, Brooks SM. Perceptions of discrimination: a multiple needs model perspective. J Manage. 2008;34(5):952–77.

Goleman D. The emotionally competent leader. Healthc Forum J. 1998;41(2):36, 38, 76.

Greene JD, Nystrom LE, Engell AD, Darley JM, Cohen JD. The neural bases of cognitive conflict and control in moral judgment. Neuron. 2004;44(2):389–400.

Gross JJ. Emotion regulation: affective, cognitive, and social consequences. Psychophysiology. 2002;39(3):281–91.

Gu X, Hof PR, Friston KJ, Fan J. Anterior insular cortex and emotional awareness. J Comp Neurol. 2013;521(15):3371–88.

Haidt J. The moral emotions. In: Davidson RJ, Scherer KR, Goldsmith HH, editors. Handbook of affective sciences. Oxford: Oxford University Press; 2003. p. 852–70.

Harmon-Jones E, Sigelman J. State anger and prefrontal brain activity: evidence that insult-related relative left-prefrontal activation is associated with experienced anger and aggression. J Pers Soc Psychol. 2001;80(5):797–803.

Harmon-Jones E, Sigelman J, Bohlig A, Harmon-Jones C. Anger, coping, and frontal cortical activity: the effect of coping potential on anger-induced left frontal activity. Cogn Emot. 2003;17(1):1–24.

Hatfield E, Cacioppo JT, Rapson RL. Emotional contagion. New York: Cambridge University Press; 1994. 240-vii p.

Hendricks L, Bore S, Aslinia D, Morriss G. The effects of anger on the brain and body. Natl Forum J Couns Addict. 2013;2(1):2–5.

Herrero N, Gadea M, Rodríguez-Alarcón G, Espert R, Salvador A. What happens when we get angry? Hormonal, cardiovascular and asymmetrical brain responses. Horm Behav. 2010;57 (3):276–83.

Holt R, den Hond F. Sapere aude. Org Stud. 2013;34(11):1587–600.

Hortensius R, Dennis JL, Harmon-Jones E. When anger leads to aggression: induction of relative left frontal cortical activity with transcranial direct current stimulation increases the anger–aggression relationship. Soc Cogn Affect Neurosci. 2011;7(3):342–7.

Jagannathan S, Rai R. Organizational wrongs, moral anger and the temporality of crisis. J Bus Ethics. 2016;141(4):709–30.

Jehn KA. A multimethod examination of the benefits and detriments of intragroup conflict. Adm Sci Q. 1995;40(2):256.

Kantor J, Twohey M. Harvey Weinstein paid off sexual harassment accusers for decades [Internet]. The New York Times. 2017 (cited 2017 Dec). Available from https://www.nytimes.com/2017/10/05/us/harvey-weinstein-harassment-allegations.html

Kassinove H. How to recognize and deal with anger [Internet]. American Psychological Association; 2017. Available from http://www.apa.org/helpcenter/recognize-anger.aspx

Kelly JR, Barsade SG. Mood and emotions in small groups and work teams. Organ Behav Hum Decis Process. 2001;86(1):99–130.

Khomami N. #MeToo: how a hashtag became a rallying cry against sexual harassment [Internet]. The Guardian. 2017 (cited 2017 Dec). Available from https://www.theguardian.com/world/2017/oct/20/women-worldwide-use-hashtag-metoo-against-sexual-harassment

Koh HC, El'Fred HY. The link between organizational ethics and job satisfaction: a study of managers in Singapore. J Bus Ethics. 2001;29(4):309–24.

Kopelman S, Rosette AS, Thompson L. The three faces of Eve: strategic displays of positive, negative, and neutral emotions in negotiations. Organ Behav Hum Decis Process. 2006;99 (1):81–101.

Lerner JS, Keltner D. Beyond valence: toward a model of emotion-specific influences on judgement and choice. Cogn Emot. 2000;14(4):473–93.

Lerner JS, Tiedens LZ. Portrait of the angry decision maker: how appraisal tendencies shape anger's influence on cognition. J Behav Decis Mak. 2006;19(2):115–37.

Lerner JS, Goldberg JH, Tetlock PE. Sober second thought: the effects of accountability, anger, and authoritarianism on attributions of responsibility. Pers Soc Psychol Bull. 1998;24(6):563–74.

Lindebaum D, Geddes D. The place and role of (moral) anger in organizational behavior studies. J Organ Behav. 2016;37(5):738–57.

Lindebaum D, Geddes D, Gabriel Y. Moral emotions and ethics in organisations: introduction to the special issue. J Bus Ethics. 2016;141(4):645–56.
Luthans F. The need for and meaning of positive organizational behavior. J Organ Behav. 2002;23(6):695–706.
Madera JM, Brent SD. The effects of leader negative emotions on evaluations of leadership in a crisis situation: the role of anger and sadness. Leadersh Q. 2009;20(2):103–14.
Martin RC, Dahlen ER. The angry cognitions scale: a new inventory for assessing cognitions in anger. J Ration Emot Cogn Behav Ther. 2007;25(3):155–73.
Mayer JD, Roberts RD, Barsade SG. Human abilities: emotional intelligence. Annu Rev Psychol. 2008;59(1):507–36.
Mills H. Physiology of anger [Internet]. Mental help. 2005 (cited 2017 Dec). Available from https://www.mentalhelp.net/articles/physiology-of-anger/
Miron-Spektor E, Rafaeli A. The effects of anger in the workplace: when, where, and why observing anger enhances or hinders performance. In: Research in personnel and human resources management. 2009. p. 153–78.
Moll J, Schulkin J. Social attachment and aversion in human moral cognition. Neurosci Biobehav Rev. 2009;33(3):456–65.
Moll J, de Oliveira-Souza R, Eslinger PJ. Morals and the human brain: a working model. Neuroreport. 2003;14(3):299–305.
Moll J, de Oliveira-Souza R, Moll FT, Ignácio FA, Bramati IE, Caparelli-Dáquer EM, et al. The moral affiliations of disgust: a functional MRI study. Cogn Behav Neurol. 2005;18(1):68–78.
Morris MW, Keltner D. How emotions work: the social functions of emotional expression in negotiations. Res Org Behav. 2000;22:1–50.
Neuman JH, Baron RA. Workplace violence and workplace aggression: evidence concerning specific forms, potential causes, and preferred targets. J Manage. 1998;24(3):391–419.
Novaco RW. Anger control: the development and evaluation of an experimental treatment. Lexington, MA: Lexington Books; 1975.
Novaco RW. Stress inoculation: a cognitive therapy for anger and its application to a case of depression. J Consult Clin Psychol. 1977;45(4):600–8.
Pawliczek CM, Derntl B, Kellermann T, Gur RC, Schneider F, Habel U. Anger under control: neural correlates of frustration as a function of trait aggression. PLoS One. 2013;8(10):e78503.
Pearson CM, Andersson LM, Wegner JW. When workers flout convention: a study of workplace incivility. Hum Relat. 2001;54(11):1387–419.
Potegal M, Stemmler G, Spielberger C. International handbook of anger: constituent and concomitant biological, psychological, and social processes. New York: Springer; 2010.
Quigley BM, Tedeschi JT. Mediating effects of blame attributions on feelings of anger. Pers Soc Psychol Bull. 1996;22(12):1280–8.
Ragins BR, Winkel DE. Gender, emotion and power in work relationships. Hum Resour Manag Rev. 2011;21(4):377–93.
Redmond V, Jameson JK, Binder A. How superior–subordinate relationship quality and conflict management styles influence an employee's use of upward dissent tactics. Negot Confl Manag Res. 2016;9(2):158–76.
Rest JR. Moral development: advances in research and theory. New York: Praeger; 1986.
Richards JM, Gross JJ. Composure at any cost? The cognitive consequences of emotion suppression. Pers Soc Psychol Bull. 1999;25(8):1033–44.
Rozin P, Lowery L, Imada S, Haidt J. The CAD triad hypothesis: a mapping between three moral emotions (contempt, anger, disgust) and three moral codes (community, autonomy, divinity). J Pers Soc Psychol. 1999;76(4):574–86.
Scherer KR. The role of culture in emotion-antecedent appraisal. J Pers Soc Psychol. 1997;73(5):902–22.
Schiff BB, Bassel C. Effects of asymmetrical hemispheric activation on approach and withdrawal responses. Neuropsychology. 1996;10(4):557–64.

Seo D, Patrick CJ, Kennealy PJ. Role of serotonin and dopamine system interactions in the neurobiology of impulsive aggression and its comorbidity with other clinical disorders. Aggress Violent Behav. 2008;13(5):383–95.

Shaver P, Schwartz J, Kirson D, O'Connor C. Emotion knowledge: further exploration of a prototype approach. J Pers Soc Psychol. 1987;52(6):1061–86.

Sims RR. The institutionalization of organizational ethics. J Bus Ethics. 1991;10(7):493–506.

Sinaceur M, Tiedens LZ. Get mad and get more than even: when and why anger expression is effective in negotiations. J Exp Soc Psychol. 2006;42(3):314–22.

Sobotka SS, Davidson RJ, Senulis JA. Anterior brain electrical asymmetries in response to reward and punishment. Electroencephalogr Clin Neurophysiol. 1992;83(4):236–47.

Spielberger CD. STAXI-2: State-trait anger expression inventory-2: Professional Manual. 1999.

Spielberger CD, Reheiser EC, Sydeman SJ. Measuring the experience, expression, and control of anger. Issues Compr Pediatr Nurs. 1995;18(3):207–32.

Srivastava K. Emotional intelligence and organizational effectiveness. Ind Psychiatry J. 2013;22 (2):97.

Stickney LT, Geddes D. Positive, proactive, and committed: the surprising connection between good citizens and expressed (vs. suppressed) anger at work. Negot Confl Manag Res. 2014;7 (4):243–64.

Sy T, Côté S. Emotional intelligence: a key ability to succeed in the matrix organization. Int J Manage Enterp Dev. 2004;23(5):437–55.

Tice B. Controlling anger: self-induced emotion change. In: Wegner P, editor. Handbook. Englewood Cliffs: Prentice-Hall; 1993.

Tripp TM, Bies RJ. What's good about revenge: the avenger's perspective. In: Lewicki RJ, Bies RJ, Sheppard BH, editors. Research on negotiation in organizations. Greenwich: JAI; 1997. p. 145–60.

Turner RJ, Russell D, Glover R, Hutto P. The social antecedents of anger proneness in young adulthood. J Health Soc Behav. 2007;48(1):68–83.

Ursin R. Serotonin and sleep. Sleep Med Rev. 2002;6(1):55–67.

Valentine S. Organizational ethics and stakeholder wellbeing in the business environment. IAP; 2014.

van Dijk E, van Kleef GA, Steinel W, van Beest I. A social functional approach to emotions in bargaining: when communicating anger pays and when it backfires. J Pers Soc Psychol. 2008;94 (4):600–14.

Van Kleef GA, Côté S. Expressing anger in conflict: when it helps and when it hurts. J Appl Psychol. 2007;92(6):1557–69.

Van Kleef GA, De Dreu CKW, Manstead ASR. The interpersonal effects of anger and happiness in negotiations. J Pers Soc Psychol. 2004;86(1):57–76.

Vytal K, Hamann S. Neuroimaging support for discrete neural correlates of basic emotions: a voxel-based meta-analysis. J Cogn Neurosci. 2010;22(12):2864–85.

Weinstein F, Stearns CZ, Stearns PN. Anger: the struggle for emotional control in America's history. J Interdiscip Hist. 1988;19(2):340.

Yao S, Ding C, Qi S, Yang D. Value associations of emotional faces can modify the anger superiority effect: behavioral and electrophysiological evidence. Soc Cogn Affect Neurosci. 2014;9(6):849–56.

Zenteno-Hidalgo A, Geddes D. Chapter 9 A model of compassionate responses to anger expression. In: Research on emotion in organizations. 2012. p. 257–77.

Workplace in Space: Space Neuroscience and Performance Management in Terrestrial Environments

15

Koji Tachibana

Contents

Abstract

I investigate how isolated and confined environment research in human space exploration informs performance management in terrestrial organizations and workplaces and to what extent space neuroscience can contribute to such research and management. Space life science studies of astronauts' performance management during long-duration missions under isolated and confined environment

K. Tachibana (✉)
Faculty of Humanities and Social Sciences, Kumamoto University, Kumamoto, Japan
e-mail: koji-t@kumamoto-u.ac.jp

J. T. Martineau, E. Racine (eds.), *Organizational Neuroethics*, Advances in Neuroethics, https://doi.org/10.1007/978-3-030-27177-0_15

started in the mid-1990s. This research merits further scrutiny, as scientific knowledge on life in space has been successfully applied to support people in related terrestrial environments. However, space neuroscientists have yet to fully explore the potential of research on the neural substrates of human behavioral health in isolated and confined environments. I demonstrate why research on these neural mechanisms can and should be conducted, identifying five principal reasons. This kind of research will contribute to facilitate the establishment of a more responsive behavioral health program in the workplace and organization, change the organizational culture, and solve issues related to international or multicultural work environments. Further neuroethical considerations are also required, since such research can be used for unethical purposes, such as augmenting workers to continue working under unhealthy workplace conditions. In conclusion, space neuroscience, in combination with other fields, can make a distinctive research contribution not merely to elucidate how the human mind, body, and brain affect ethical performance in various terrestrial, isolated, and confined settings but also to improve performance management and establish ethically sound organizations.

15.1 Ethical Instability in Isolated and Confined Environments

In any workplace where managers and employees work together, there is inevitably some risk of unethical behavior that colleagues and even the person behaving unethically may not predict beforehand. For a period of about 10 years between the late 1990s and the early 2000s, a series of incidents illustrating this risk occurred in the USA, now known as the "strip search phone call scam." A man called a fast-food restaurant or grocery store in a small town in a rural area, claiming to be a police officer. He then convinced managers to conduct strip searches of female employees, as well as other strange acts, in isolated and confined rooms, such as offices or restrooms, claiming that this was on behalf of the police. This series of incidents shocked US citizens, as it emerged that, in isolated and confined conditions, even good citizens would readily behave unethically under the supposed authority of the police or other such agencies. The suggestion to be drawn from this series of incidents is that workplaces may be ethically unstable.

On reflection, however, the ethical instability of ordinary people is not so surprising. Since the 1960s, social psychologists have reported evidence that, under isolated and confined conditions, we may have difficulty in maintaining ethically sound emotional, cognitive, and behavioral responses and rather commit wrongdoings. The obedience experiment that Stanley Milgram and his colleagues conducted at Yale University is one of the well-known studies on this phenomenon (Milgram 1963, 1974). They showed that citizens who were taken to an underground experimental laboratory readily acceded to the authority of an experimenter wearing a gray lab coat and administered severe electric shocks as a punishment to another (pseudo) participant, following the experimenter's instructions. In fact, against the researchers' expectations, 25 of the 40 participants (62.5%) were willing to administer the most powerful shock, which the subject believed to be 450 V. Following the experiment, some participants developed psychiatric symptoms similar to post-traumatic stress disorder (Blass 2004). The experiment came to be known as the Eichmann experiment, because it mimicked the actions of Adolf Eichmann, the architect of the Holocaust during the Nazi regime. Having carried out his duties in a meticulous and even rigid manner, he explained at his subsequent trial in 1961 that "I'm a normal person, and I was just following orders from above," now known as the banality of evil (Arendt 1963). Another well-known study is the so-called Stanford prison experiment that Philip Zimbardo and his colleagues conducted at Stanford University (Haney et al. 1973). In this experiment, 24 male participants, who were judged to be socially, mentally, and physically healthy, were randomly divided into guards and prisoners. Inhabiting a prisonlike environment, they became highly involved in their roles to an extent that was contrary to and well beyond the researchers' expectations. The "guards" behaved in a heavy-handed, sadistic manner, befitting their role, oppressing the "prisoners," and undermining their dignity. In response, the prisoners rebelled just 2 days into the experiment, but the guards suppressed the rebellion, and as they even enforced heavier punishment, some of the prisoners became mentally disturbed. As a result, the experiment was terminated on the 6th day. Both experiments suggest that ordinary "good" people, who were mentally healthy and had never caused any social problems, had difficulty in maintaining ethically sound character and, under isolated and confined conditions, behaved in unethical ways.

These studies concern not merely psychological issues but also raise moral questions (Benjamin and Simpson 2009). From a philosophical point of view, the results of these studies offer important insights into the nature of human ethical character. In fact, they contribute to recent philosophical debate on the nature, where psychology-minded moral philosophers argue that traditional virtue ethics is wrong, because its central thesis is empirically inadequate: the central thesis is that human morality consists in a cross-situational stable ethical character, but those psychological studies show that even good people do not have such a stable character (Doris 2002; Doris and the Moral Psychology Research Group 2010; cf. Sreenivasan 2013). These psychological studies contribute to elucidating how human ethical character fails to demonstrate its performance—an amalgam of ethically sound emotional responses, cognitive states, and behaviors in social and interpersonal situations—in isolated and confined environments. From an everyday practical perspective, the better we understand this aspect of human ethical nature, the better we can consider how an organization can maintain an ethically sound working environment, where employees do not commit wrongdoings, such as harassment and embezzlement. Then, we may be able to develop an ethically sounder performance management practice in the workplace. Therefore, further research will contribute in accomplishing these ends.

However, as a matter of fact, we have more difficulty in finding recent sociopsychological studies of this kind, except a few (Burger 2009; Simpson et al. 1992). The scarcity of such studies is also observed in the abovementioned debate on the fragility of human ethical character, where almost all studies considered are the early ones between the 1960s and 1970s.

Two problems can be identified in this respect. The first problem is the development of research ethics regulations. Although some of the early studies violated even the standards of that time, studies of this kind are all the more problematic in light of current research ethics standards (Blass 2004; Zimbardo 2007; Benjamin and Simpson 2009). The second problem is the poor reproducibility of research, that is reported as a general issue in psychology (Open Science Collaboration 2015). In fact, not a small number of psychologists cast a doubt also on the validity of Zimbardo's sociopsychological experiment (Blum 2018; Gray 2013; Resnick 2018; Le Texier 2019; cf. Carnahan and McFarland 2007). These two problems may prevent psychologists from conducting further empirical studies on human character in isolated and confined environment.

However, the terrestrial environment is not the only option for conducting further research on human psychology and behavior in isolation and confinement, nor is such further research limited to the field of social psychology. Medical and psychological research on the behavioral health of astronauts in space can be another option. Using the disciplines of medicine, psychology, neuroscience, and behavioral science, this research field can now inherit the research topic and explore human nature in space environment. Focusing on isolated and confined environment research in human space exploration, I will examine how space neuroscience may contribute to the development and enhancement of performance management practices in terrestrial environments, such as organizations and workplaces.

15.2 Advances in Space Exploration, Performance Management in Space, and Terrestrial Applications

Space exploration has always aimed at going further and staying longer in outer space. In particular, the so-called Space Race between the USSR and the USA, which began in the 1950s, led to a period of rapid and accelerating development in both unmanned and human space explorations. Examples of unmanned space exploration include the launch of an artificial satellite (Sputnik 1, 1957) and missions to reach and document the lunar surface (Lunar 2 and Lunar 3, 1959). The human space explorations include Gagarin's 108-minute space flight (Vostok 1, 1961) and Neil Armstrong and Buzz Aldrin's lunar landing (Apollo 11, 1969). Since the 1970s, unmanned artificial satellites have been launched with the aim of exploring more distant planets, such as Mars and Jupiter, while astronauts have completed longer missions in space, with Soviet cosmonauts in the *Mir* space station repeatedly breaking human mission duration records.

Since the 1990s, in the period following the end of the Cold War, a new regime of space exploration has been observed, namely, the gradual evolution from the hotly contested Space Race into the commitment to international cooperation. In the field of unmanned space exploration, for example, a joint initiative of the National Aeronautics and Space Administration (NASA) and the European Space Agency (ESA) sent a spacecraft named Cassini–Huygens to Saturn, thus reaching the planet's moon, Titan, in 2005. In the field of human space exploration, the International Space Station (ISS) has been in operation since 1998 as an initiative of fifteen countries, including Russia, the USA, Canada, Japan, and some of the member countries of the ESA. This ongoing international cooperation enables multicultural astronauts to spend 6 months in average together, undertaking a range of precisely scheduled missions.

Beyond their significance for the evolution of internationally cooperative space exploration in the twenty-first century, those long-duration missions highlight space-medical considerations concerning the health and performance management of astronauts in space. The exceptional competence of astronauts cannot be denied; they must undergo rigorous screening and training, which demands physical toughness, mental stability, intellectual brilliance, and social extroversion. Yet, despite meeting these high standards, even they occasionally fail to demonstrate such competences, as a result of the multiple stressors that characterize the space environment, including microgravity, weightlessness, space radiation, changes in circadian rhythms (i.e., the 45-min day-night cycle), as well as the isolated and confined environment (Barratt and Pool 2008; Clément 2011). The longer astronauts stay in space, the more severe the influence of these stressors is likely to be in terms of performance management. For example, in relation to the physiological influences of microgravity, it is reported that a 6-month stay in the *Mir* space station causes the volume of the plantar flexor muscle in the calf to decrease by 6–20%, while muscle tension (maximum voluntary contraction force) likely decreases by 20–48% (Zange et al. 1997). More recently, it was reported that, when compared to short-duration missions, such as space shuttle flights, astronauts on long-duration missions in the ISS experience visual impairments, which are partly irreversible due to volume changes in the cerebrospinal fluid caused by microgravity (Alperin et al. 2017). Space radiation, another stressor, is reported to potentially affect the vascular endothelium, thus causing Apollo lunar astronauts to suffer 4–5 times higher cardiovascular disease mortality than non-flight and low Earth orbit astronauts (Delp et al. 2016). As one of the effects of weightlessness, 75% of astronauts suffer from space motion sickness on their maiden flight (Young 2000). Noise in the ISS reaches an average of 70 dB (with a maximum exceeding 90 dB), thus interrupting astronauts' sleep up to 5–8 times per night (Mallis and DeRoshia 2005).

While space-medical research on these and other physiological stressors is well established, the research on those that weaken the behavioral health of astronauts and deteriorate their psychosocial performance is a relatively new topic, which has been studied only since the mid-1990s (Harrison and Fiedler 2013a). Still, space medicine plays a central role not only in physiological but also in behavioral aspects of astronauts' healthcare (Clément 2011; Nicogossian et al. 2016). In fact, in collaboration with related studies that focus on analogous environments on Earth, such as submarines, Antarctic expedition bases, and experimental facilities, space medicine continuously progresses (Baranov 2001; Bishop 2013; Bonting 1993; Harrison et al. 1991). Current evidence suggests that behavioral health and performance of astronauts are mainly influenced by psychological and interpersonal stressors arising from long-duration missions with multicultural crews under conditions of isolation and confinement (Flynn 2008; Kanas and Manzey 2008). Several articles have reported that the workplace of astronauts in space, namely, the ISS, undermines the behavioral health and performance of crew members, thus resulting in inappropriate subgroup behavior, disordered leadership, cultural misunderstandings, harassment, quarrels, dropouts, and symptoms of psychiatric disorders, including depression, anxiety, and asthenia (Bessone et al. 2008a, b; Flynn 2008; Kanas and Manzey 2008; Tafforin 2013). At present, NASA gives a list of predictors and contributing factors concerning the behavioral health of astronauts: (1) personality, (2) resiliency and hardiness, (3) emotional reaction, (4) sleep and circadian rhythm, (5) habitability and environmental design, (6) job design (autonomy and meaningful work), (7) monotony and boredom, (8) daily hassles and major life events, (9) cultural factors, (10) ground mission support, (11) family and social support, and (12) world events (Slack et al. 2016).

Behavioral health, which space medicine and psychology have studied since the mid-1990s in terms of the performance management, can be characterized as a wider concept than earlier formulations, such as mental health, in that "it recognizes that effective, positive behavior depends on an interaction with the physical and social environments, as well as an absence of neuropsychiatric dysfunction. Behavioral health is evident not only at the level of the individual, but also at the levels of the group and organization" (Harrison 2005: B3). Therefore, such behavioral health is important in terms not only of the astronaut's psychosocial adaptation but also of their both individual and group performance (Harrison and Fiedler 2013b) and even of their well-being (Bishop 2013), especially in the long-duration space missions, such as the ISS missions and human Mars exploration. From a philosophical point of view, such performance can be understood as *ethical* performance. The word "ethical," different from "moral," is originally an evaluative word that covers a variety of personality-derived human responses, including emotional, cognitive, and behavioral responses, in both social and interpersonal contexts, and from both temporal and long-term perspectives (Aristotle 1998). Therefore, we could say that medical and psychological research on behavioral health of astronauts has investigated the mechanism and management of their ethically sound performance.

The link between behavioral health and ethical performance suggests a terrestrial application of medical and psychological knowledge on astronauts' behavioral healthcare in space. So far, such knowledge has been successfully applied to supporting ordinary people in terrestrial isolated and confined environments on two occasions (Tachibana et al. 2017). The first such case occurred in Japan in 2009, when 48 passengers and crew members on flight NWA25 from the USA were found to be infected with a new strain of influenza (A/H1N1), which had emerged in Mexico, spread across the world. Since they were quarantined by law in an isolated and confined facility in Narita City, Japan's Ministry of Health, Labour, and Welfare requested the assistance of Colonel Dr. Tachibana (MD) of Japan Aerospace Exploration Agency (JAXA), who had served as the flight surgeon and had been providing psychiatric support to Japanese astronauts stationed on the ISS. Dr. Tachibana and his colleagues applied their accumulated space-medical knowledge on astronauts' behavioral healthcare to support the passengers, crew, and quarantined airport staff.

They subsequently reported that "[f]rom the experiences of providing comprehensive health support for astronauts stationed in the isolated and confined environment of the ISS, it should be possible to provide mental health support to the people subjected to quarantine measures" (Tachibana et al. 2012: 31). The second case occurred in Chile in 2010, now widely known as the Chilean Mining Accident, in which 33 male mine workers were trapped 700 m underground for 69 days. In response to the demand of the Chilean government, NASA dispatched four members of the ISS' astronaut health management team to Chile to provide advice and direction for the rescue effort, including the creation of a team to ensure the psychological and behavioral stability of the trapped miners, while at the same time "assessing the parallels between the miner's plight and life in space" (NASA 2011). Dr. Duncan (MD), NASA's deputy chief medical officer and leader of this team, reported that "we were able to bring the knowledge we learned in space to the surface and under the surface, in order to help people here on Earth" (NASA 2011).

The above two cases suggest that medical and psychological knowledge on astronauts' behavioral healthcare in space can have a wider terrestrial application, such as the provision of behavioral healthcare support for other groups in isolated and confined environments on Earth, including evacuation sites, schools, and workplaces. Recently, a list of factors is presented, aiming to utilize such knowledge in assessing ethical performance on terrestrial environments, as follows: (1) time in isolation, (2) circumstances in which the situation arises, (3) preparation and training, (4) threat to life, (5) methods of access to information and communication, (6) role of all individuals, (7) group unity, (8) methods of eliminating stress, (9) challenges in everyday life, (10) external support, (11) reaction and support from society, and (12) mental state at time of release (Tachibana et al. 2017; see also Bower et al. 2019). This suggests that space medicine and psychology provide knowledge on behavioral health that is not local and limited, applicable only to space environment, but rather knowledge that contributes to understanding human ethical nature in general and accordingly to supporting ordinary people on Earth (Tachibana et al. 2017).

15.3 Neuroscience in Space: Its History and Prospects

Space medicine is a focal science in the emerging research network of space life sciences, encompassing space physiology, space psychology, space neuroscience, and other related research fields (Buckey 2006; Clément and Reschke 2008; De la Torre 2014). Accordingly, neuroscience research on both physiological and behavioral aspects of the astronauts' health in space must be understood within the context of this research network.

The first neuroscience experiments in space were conducted during the Soviet Vostok-3 mission in 1962, when crewman Andriyan Nikolayev completed a number of sensory-motor coordination tests based on electroencephalography (EEG), electrooculography (EOG), and galvanic skin response (GSR) (Clément and Reschke 2008). Since then, both the USSR and the USA have conducted a wide range of experiments related to human otolith function, in-flight sleep, visual acuity, vestibular function, motor sensory performance, and locomotion, monitored with EEG. However, few studies looked at behavioral health at that time. In 1977, for example, two Russian cosmonauts, Yury Romanenko and Georgi Grechko, who stayed on Soyuz-26/Salyut-6 for more than 3 months, conducted many experiments, including the first neuroscientific experiment on space motion sickness (SMS), with the use of EEG. SMS is closely related to behavioral health since both concern astronaut performance. However, experiments on behavioral health had not yet been studied in space at that time. It was in 1988, on Soyuz-TM5/Mir-3, that the first neuroscientific study concerning the behavioral health of astronauts in space was conducted, where Russian cosmonauts carried out psychomotor tests.

In the 1990s, neuroscience came in the spotlight, as the US Federal Government and the National Congress designated this decade as the Decade of the Brain and promoted neuroscience research on human brain. Furthermore, it was also the decade when the importance of research on astronauts' behavioral health under long-term isolation and confinement gradually came to be appreciated by space life scientists (Ursin et al. 1991). However, even in the 1990s, neuroscientific experiments on behavioral health in space were rare. For example, in the early 1990s, it was anticipated that, in the 2000s, Space Station Freedom—later converted into the ISS—and other long-duration missions would provide opportunities to explore new directions in neuroscience, such as sensory substitution, augmentation, and robotic telepresence; however, no reference was made to the neural basis of the ethical performance deterioration in long-duration missions (Sulzman and Wolfe 1991). In fact, as part of the national policy for the Decade of the Brain, NASA conducted a series of shuttle/spacelab missions, named *Neurolab*, in the 1990s, in order to research the nervous system and behavior. Nine crews conducted 26 neuroscience experiments on various topics concerning balance, sensory integration and regulation, nervous system development in weightlessness, and blood pressure control, but no experiments were conducted on behavioral health (Buckey and Homick 2003; NASA 2000).

It was presumably in the twenty-first century that the next related neuroscientific study was conducted in space; in 2006 on Soyuz-TMA9 (ISS Expedition-14), crews investigated "individual features of state psychophysiological regulation and crewmembers' professional activities during long space flights" (Clément and Reschke 2008: 69). Neuroscientific research on the behavioral health of astronauts in long-duration missions under isolated and confined environments gained recognition little by little. For example, among the neuroscientific studies on human and nonhuman physical and cognitive abilities conducted to date during more than 200 missions, behavioral health is included as one of the major research topics in space neuroscience, as follows: (1) operational aspects, such as space motion sickness, sleep disorder, and psychosocial-behavioral performance; (2) sensory functions such as vision, hearing, taste, olfaction, proprioception, and the sense of motion; (3) pre-/in-/post-flight posture, movement, and locomotion; (4) compensatory eye movements; and (5) spatial orientation (Clément and Reschke 2008). However, even in the era of very long-duration space flights where decrements in operational task performance have been observed, "only a few studies were performed" on the behavioral health-related astronaut performance (Clément and Reschke 2008: 94; see also Clark and Bacal 2008; De la Torre 2014).

In the future, space neuroscience can develop in at least three ways. The first is the most promising, namely, to identify the neural mechanisms by which space environment, including space radiation, microgravity, and weightlessness, influences the human body and physical and cognitive performances. For example, a recent article reports that space radiation may alter synaptic activity in the central nervous system, causing cognitive performance deficits and leading to long-term neurocognitive sequelae (Parihar et al. 2015). Using magnetic resonance imaging (MRI), another study reports that narrowing of the central sulcus, upward shift of the brain, and narrowing of cerebrospinal fluid spaces at the vertex were observed among the consequences of long-duration missions (Roberts et al. 2017). EEG research using parabolic flights is also conducted to study the influence of weightlessness on the brain (Schneider et al. 2008). More recently, a study reports the recovery process of neurocognitive functions of astronauts after flight, by using functional MRI (Seidler et al. 2018).

The second is to apply brain computer interface (BCI) and brain machine interface (BMI) techniques in space environments. Due to the hostility and danger of space environments for human beings, BCI-/BMI-based robotic devices are becoming increasingly useful for completing risky tasks and missions involving extravehicular activity (EVA) (Millán et al. 2007). However, technical and ethical considerations should be noted about this prospect. As a technical note, BCI/BMI combined with EEG or near-infrared spectroscopy (NIRS) currently has technical limitations, even if these devices are the most promising options in space (Coffey et al. 2010). Regarding ethical terms, neuroethical and space-ethical considerations will be required about enhancing astronauts, given that BCI/BMI devices for astronauts can be viewed as a form of augmentation or enhancement rather than as health supports (Citi et al. 2009).

The third is an interdisciplinary investigation of the neural basis of astronauts' behavioral health in space. Some neuroscientists have proposed as a future research program the "identification of neural mechanisms, neurotransmitters, and neural substrates that impact on optimal psychological performance, especially in areas of mood, well-being, motivation, and reward" (Clément and Reschke 2008: 96). The advancement of this research will provide appropriate countermeasures and support systems for astronauts' behavioral health and accordingly contribute in avoiding more severe situations (Kanas et al. 2009). Up to now, in the interest of behavioral healthcare, astronauts periodically conduct cognitive task assessments, and flight surgeons have structured interviews with them (Flynn 2008). However, insofar as it concerns ethical performance, research on astronauts' behavioral health needs to be interdisciplinary (Harrison and Fiedler 2013b). Many approaches fall under the umbrella of this research, including the analysis of astronauts' journals (Stuster 2016), communications with ground control (Shved et al. 2014), sociopsychological analysis (Kuznetsova et al. 2017), ethological and anthropological analysis (Tafforin and Abati 2017), facial expression and computer analysis of speech (Harrison and Fiedler 2013b), physiological responses (Vinokhodova et al. 2012), and even virtual reality (Salamon et al. 2018). Neuroscientific approaches also fall under this umbrella. At present, there are few terrestrial studies concerning behavioral health, such as the one on the relationship between exercise and behavioral health under terrestrial experimental space chamber, using EEG (Kovaleva et al. 2017; Schneider et al. 2010). In combination with those and other approaches, neuroscience in space can study not only physiological or cognitive but also behavioral health and ethical performance. Although yet to play a central role in space life science research, neuroscientific methods seem worthy of further development.

15.4 Five Reasons for Advancing Space Neuroscience Research on Human Ethical Performance

The previous section showed that neuroscience research on human ethical performance in space has yet to receive enough attention. This section will show that at least five reasons can be identified for paying attention to and advancing such neuroscience research *in space*.

15.4.1 Unique Characteristics of the ISS as a Research Setting

A hallmark of space neuroscience on ethical performance is that the ISS itself provides a unique research setting for examining situation-laden human ethical nature. As mentioned above, the ISS is an isolated and confined environment where astronauts with multicultural backgrounds stay together for 6 months in average. Space medicine studies various stresses caused by environments that undermine the crew's performance. However, such studies are not utterly unique but are complemented by studies on analogous terrestrial environments, such as experimental space chambers on Earth (Baranov 2001; Bonting 1993), Antarctic expedition bases and submarines (Harrison et al. 1991; Bishop 2013), and quarantine facilities (Abad et al. 2010; Jeong et al. 2016; Maunder et al. 2003; Tachibana et al. 2012), which have shown that human ethical and cognitive performances deteriorate under isolation or confinement.

The uniqueness of the ISS' environment lies in the fact that the ISS is a completely human-made facility, in which every parameter necessary for human survival is controlled, monitored, and collected by a range of professionals on a 24/7 basis. The Environmental Control and Life Support System (ECLSS) manages the oxygen, nitrogen, and carbon dioxide contents of the air, the noise level of working and living areas, temperature, humidity, atmosphere, water, food, waste, and so on (NASA 2008, 2017). Professional officers at ground stations, such as the flight director (FD), capsule communicator (CAPCOM), and flight surgeon (FS), not only support missions but also monitor and collect data on the physicochemical, biological, cognitive, psychological, psychiatric, and behavioral phases. Therefore, the ISS' research setting will allow the investigation of the extent to which each factor influences behavioral health and performance, the correlations between them, and accordingly the conditions under which human sociomoral performance deteriorates and can be fully displayed. This is a unique advantage compared to studies using terrestrial isolated and confined environments, since terrestrial research cannot control or monitor every factor in the environment that can influence target performance.

15.4.2 External and Ecological Validities

The first advantage makes neuroscience research in space valid in two ways. As an experimental science, neuroscience must ground its arguments on experimental evidence. Related experiments must satisfy a number of conditions: they use experimental devices, such as EEG, NIRS, and functional MRI; they are based on concepts like "operational definition," "manipulation," and "control"; and they are modulated by statistical significance detection and controlled variables. These internal conditions are necessary, though not sufficient, for well-designed experiments, since a well-designed experiment also needs to be externally and ecologically valid (Tachibana 2009).

The concepts of external and ecological validities have been discussed in many disciplines adjacent to neuroscience, highlighting how they differ from each other (Brewer and Crano 2014; Franzen and Wilhelm 1996; Mitchell and Jolley 2004; Sbordone 1996; Shadish et al. 2002; see also Houts and Gholson 1989). In brief, external validity indicates whether an experimental result (and the experimental design) can be generalized to other situations such as different populations, settings, and so on, whereas ecological validity indicates whether these situations adequately represent the real-world setting. However, there is no consensus regarding the status of ecological validity and its relationship with external validity. For example, some argue that ecological validity is not a type of validity but rather a research method or a process for reducing uncertainty (Franzen and Wilhelm 1996; Shadish et al. 2002); others say that it is an important form or face of external validity (Brewer and Crano 2014). In a slightly different context, it is suggested that ecological validity "has often been ignored by neuropsychologists, particularly those who seem more interested in localizing brain damage" (Sbordone 1996: 17). But still, researchers agree that tasks, situations, and results of an experiment must represent the real-world settings as much as possible because the *explanandum* is a phenomenon in the real world. Therefore, the more externally and/or ecologically valid an experimental design is, the better the experiment is.

The ISS can offer both validities. External validity of space medicine and psychology research will be enhanced by the fact that crew members, their gender, cultural backgrounds, and missions are replaced every 3 months. Its ecological validity will be enhanced by the fact that they live and work together in reality for 6 months on average.

15.4.3 Sufficient Onboard Equipment

In general, the technical and engineering evolution and the subsequent enrichment of onboard space station equipment have been essential in advancing neuroscience studies in space (Bois et al. 1998). Neuroscientific experiments could not have been performed if appropriate equipment had not been onboard. At present, the ISS has sufficient equipment to conduct neuroscientific studies on both humans and animals. Regarding animal experiments, for example, small centrifuges make it possible to turn gravity on and off (from near 0 to 2G), which is not allowed for human subjects (Clément and Reschke 2008). Regarding human research, the ESA's chamber, *Columbus*, has both multielectrode and portable EEGs (International Space Life Science Working Group 2014). EEG is the best neuroimaging tool in the ISS, since other tools, such as functional MRI, cannot be installed because of payload restrictions and the cost of space missions (De la Torre 2014). As portable EEG devices are available for using outside laboratory settings, they can help realizing more ecologically valid experimentation in the ISS. In this way, research on the neural mechanisms of human ethical performance in isolated and confined environments can be conducted in the ISS.

15.4.4 Replication and Reproducibility

By virtue of the first three advantages, it becomes possible to conduct continuous replications of space neuroscience research that can supply rich and reliable data on human ethical nature. This will contribute to improve the reproducibility of the target research, which has been a crux of psychology research on human ethical nature in isolation and confinement.

15.4.5 Research-Ethical Soundness

The other crux is research ethics regulation. On one hand, the understanding of human ethical nature under isolated and confined environments is important. On the other hand, as described in Sect. 15.1, the current circumstances, related to research ethics, delay (or even stop) the advancement of such understanding. Since science should comply with research ethics guidelines, scientists need to find a new, ethically approved research settings for this study. The ISS offers a legitimate site for ecologically valid research on isolated and confined environments, where human beings live and work together for extended periods of time. Scientists can conduct studies on human performance in an ethically acceptable way, so long as their experiment proposals pass the two-step review process for conducting human subject research in the ISS, provided by the institutional review board (IRB) of each space agency and the human research multilateral review board (HRMRB).

These five features of the ISS enable neuroscientists to conduct unique neuroscientific experiments. In combination with other fields, such studies allow the collection of neural, physicochemical, biological, psychological, and behavioral data on the maintenance and deterioration of human ethical performance in an isolated and confined environment. Exploiting the potential to collect such multilayered data, space neuroscience can investigate the influence of isolated and confined environment on human ethical performance at various levels in externally and ecologically valid ways. Furthermore, such studies can satisfy research ethics standards and be conducted to assess their reproducibility. These reasons indicate the distinctive research contribution of space neuroscience toward the elucidation of mechanisms by which the human mind, body, and brain affect both the maintenance and deterioration of ethical performance in isolated and confined environments. In this sense, space neuroscientific research on human ethical performance in such environments can be regarded as research on human ethical nature and, accordingly, as the successor of the series of sociopsychological experiments conducted by Stanley Milgram and Philip Zimbardo between the 1960s and 1970s.

15.5 The Future: Application to Terrestrial Workplaces and Possible Threats

Some astronauts and space officers have published books that offer useful guidance for business and organization management, including leadership (Searfoss 2016; Yamaguchi 2014), stress control (Furukawa 2013), and social advancement (Aldrin and Abraham 2016; Hadfield 2013; Wakata 2016). The publicity of these and other related books suggests a potential demand for astronaut's performance management in terrestrial and everyday applications. As mentioned in Sect. 15.2, astronauts undergo rigorous screening and training and are thought to have exceptional capabilities, including physical toughness, mental stability, intellectual brilliance, and social extroversion, which constitute their high-level ethical or sociomoral performance. Thus, their words are convincing to readers. However, these books are, as such, anecdotal in that the descriptions are based on personal experience to some extent (see also Wotring 2012: 2). Anecdotal reports are observed not only in astronauts' performance management but also in every field of human space exploration and have undoubtedly contributed to the progress of space sciences. Still, space life scientists call for systematic and scientific research on performance management in various fields, including medicine (Clément 2011), psychology (Kanas and Manzey 2008), and pharmacology (Wotring 2012). Such studies are expected to provide us with firmer knowledge on human performance management on Earth as well as in space. Space neuroscience could also meet this expectation if it studied the neural mechanisms of human ethical performance under isolated and confined conditions in collaboration with other related research fields within the space life science network. I propose three possible contributions of space neuroscience research as well as a threat that it might pose.

15.5.1 Establishing a Responsive Occupational Behavioral Health Program in the Workplace

Recent neuroscience research on stress has shown that the tripartite corticolimbic circuit (comprising the amygdala, hippocampus, and prefrontal cortex) operates across species to regulate both the immediate response to stress and its long-term impact (Hariri and Holmes 2015). With the identification of genetic, neural, physiological, psychological, behavioral, and environmental factors in mental and psychiatric disorders, studies on multidimensional biomarkers can inform stress assessment. In the future, preventive behavioral health programs could be based on these biomarkers, using non-pharmaceutical and noninvasive devices, such as EEG. Since such devices can be used without the supervision of a medical doctor, preventive behavioral healthcare could be conducted at the workplace. Preliminary research could readily be performed in the ISS by astronauts, by conducting EEG-based self-monitoring of their behavioral health under isolated and confined conditions. The accumulated evidence could be applied to terrestrial workplaces to inform ethically sound performance management and organizations' behavioral health programs.

15.5.2 Changing Organizational Culture

Many organizations offer educational seminars, like in-company training and career training, which aim in enhancing ethical performance of managers and employees, so that they can work effectively and avoid ethical misconducts, such as harassment and embezzlement. Such seminars would be more useful if they were based on the scientific understanding of human ethical nature, i.e., that even good and well-educated people can easily cross their ethical boundaries in a certain environment. Therefore, given that situational factors, like isolated and confined environments, inevitably deteriorate ethical performance, organizations may benefit from establishing new workplace regulations, based on that understanding, rather than advancing corporate ethics education to enhance employees' work performance straightforwardly (Alzola 2008; Salvador and Folger 2009). For example, they can inform workers about the weakness of human ethical nature, i.e., the vulnerability to certain environmental factors, such as isolation and confinement, and establish a new sort of organizational culture, based on the awareness of such factors. This sort of turnabout will be promising; if informed in advance on relevant psychological aspects, such as performance deterioration under certain situations, people can deal with ethical issues and maintain their ethical performance in these situations (Hartman 2013). Therefore, evidence on human ethical nature under isolated and confined environment would enlighten us about the rules, regulations, and working environments that are ethically sound and profitable for organizations from a neuroscience viewpoint.

15.5.3 Solving Ethical Issues Related to International or Multicultural Background

As is described in Sect. 15.2, a multicultural environment can be a stressor for a cohesive workplace in space. Since such an environment is inevitable in the ISS, space life sciences have been working on this topic and have accumulated countermeasures (Bessone et al. 2008a, b; Slack et al. 2016). Space neuroscience can contribute in understanding such stress at the neural level and aid in developing appropriate countermeasures. This knowledge can also be applied to analogous terrestrial organizations, such as workplaces, hospitals, schools, and councils. Improving the relationships among international members in industry, economy, politics, culture, education, and so on, space neuroscience will help us in establishing sound, high-performance workplaces, organizations, and collaborations.

15.5.4 The Risk of Dual Use

Despite the abovementioned contributions, further and cautious neuroethical or space-ethical considerations will be required, as space neuroscience research on this topic contains the risk of dual use or misuse. Space exploration has historically and politically been linked to military and national security. A telling example of this link is a public speech by J. F. Kennedy in the 1960 Democratic primary elections, where he reiterated the missile gap between the USSR and the USA and stressed that space exploration was a vital security issue (Tachibana et al. 2017). In fact, the [Defense] Advanced Research Projects Agency ([D]ARPA) has been involved in several military experiments in both unmanned and human space missions, including the launch of information gathering satellites (IGS) for military purposes and a laser experiment between Earth and the Moon by Neil Armstrong and Buzz Aldrin on Apollo 11 (Jacobsen 2015). Space life sciences cannot be indifferent to such military purposes. From a historical point of view, space medicine, for example, has emerged from aerospace or aviation medicine, which rapidly developed in the 1910s, in terms of combat pilot selection and flight safety, as most aircraft accidents at that time were due to human failure, especially physical defects, rather than combat (Dille and Mohler 2008). Aerospace medicine studies how different factors, to which the pilot is subject in the cockpit, including acceleration, decompression, vibration, noise, and stress, affect pilot performance, as well as the personal characteristics required to complete a mission accurately and safely. Space medicine exploits such studies when selecting astronauts and supports human space flights (Naval Aerospace Medical Institute 1991; see also Wolfe 1979).

The accumulated knowledge of such medical research is available not only for maintaining the pilot's performance but also for enhancing it. For example, it may allow producing an augmented pilot who can concentrate on his task for over 48 h or an enhanced soldier who can keep sane and accomplish long-duration missions under extraordinary isolated and confined situations. These possibilities could attract military and national security agencies, as they fit with current projects of these agencies. Since the 1960s, the USA, the UK, and German Air Forces have distributed some drugs, such as amphetamines, dexedrine, and modafinil, to combat pilots so that they could accomplish the lengthy flight missions (Mehlman 2015). The [D]ARPA examines how human physical, physiological, cognitive, and mental performance can be manipulated by neuroscientific approaches (Jacobsen 2015). Similarly, in order to capture spies and extract their confessions by psychological tortures, the Central Intelligence Agency (CIA) "studied" the influence of isolated and confined environments on human psychology (McCoy 2006). Therefore, performance management studies by space life sciences has a potential link to military and national security purposes.

Space neuroscience must be cautious with that latent link, insofar as it is a part of space life sciences. An unwilling result may be brought by mobilizing space neuroscience to inform management practices. Once knowledge of space neuroscience is applied for producing augmented soldiers, fearless in the extreme conditions of the battlefield, such military application would be within an inch of being extended further, for example, in ensuring that personnel remains functional even in unsound workplaces. Alternatively, it might also be used to lead an unpleasant employee to resign by making him/her mentally ill. The potential dual use of space neuroscience needs to be addressed by space neuroethics, by performing neuroethical examinations on the possible outcomes of space neuroscience (Tachibana 2017). Further cautious space neuroethical examinations will be required for assessing the terrestrial applications of space neuroscientific knowledge concerning human ethical performance on social organizations.

15.6 Conclusion

Space neuroscience is now more than 50 years old. The lengthening of space missions has changed the faces and foci of space life science research. Human performance management during long-duration missions in isolated and confined environments has only just started to be explored, while space neuroscience research on human behavioral health in such environments has yet to receive enough attention, even from space life scientists. However, as I demonstrated, space neuroscientists have several reasons to advance this research. As a totally human-made, isolated and confined environment, the ISS can support both externally and ecologically valid experimentation to investigate the neural substrates of human ethical performance in a research-ethically acceptable way. By virtue of the completely artificial nature of the ISS, such experiments can assemble interdisciplinary data that will illuminate the neural, physiological, and psychological mechanisms of human ethical performance. In combination with other fields and by exploiting this unique setting to collect multilayered data, space neuroscience can elucidate how the human mind, body, and brain shape performance in an isolated and confined environment. The results of this research would be mostly applicable to organizations located in relatively isolated and confined environments, such as hospitals, schools, prisons, government offices, banks, and similar private workplaces. It could contribute in maintaining the ethical performance of workers by helping to establish ethically sound working environments. Thus, space neuroscience can be the frontier of organizational neuroscience. Nevertheless, since such application may bring unwelcomed results, including the augmentation of workers under poor work environments, delicate space neuroethical discussions must also be required.

References

Abad C, Fearday A, Safdar N. Adverse effects of isolation in hospitalized patients: a systematic review. J Hosp Infect. 2010;76:97–102.

Aldrin B, Abraham K. No dream is too high: life lessons from a man who walked on the moon. Washington, DC: National Geographic; 2016.

Alperin N, Bagci AM, Oliu CJ, Lee SH, Lam BL. Role of cerebrospinal fluid in spaceflight-induced ocular changes and visual impairment in astronauts. Radiology. 2017;285(3):1063.

Alzola M. Character and environment: the status of virtues in organizations. J Bus Ethics. 2008;78:343–57.

Arendt H. Eichmann in Jerusalem: a report on the banality of evil. New York: Viking; 1963.

Aristotle. Nicomachean ethics (translated by Ross WD, revised by Ackrill JL and Urmson JO). Oxford: Oxford University Press; 1998.

Baranov VM, editor. Simulation of extended isolation: advances and problems. Firm SLOVO Moscow; 2001.

Barratt MR, Pool SL, editors. Principles of clinical medicine for space flight. New York: Springer; 2008.

Benjamin LT, Simpson JA. The power of the situation: the impact of Milgram's obedience studies on personality and social psychology. Am Psychol. 2009;64(1):12–9.

Bessone L, Coffey E, Filippova N, Greenberg E, Inoue N, Gittens M, Mukai C, Onufrienko Y, Tomi L, Schmidt L, Shea C, Shevchenko O, Sipes W, Vander Ark S, Vassin A. International space station human behavior and performance competency model, volume II. Mission Operations Directorate (MOD) International Training Control Board (ITCB) HBP Training Working Group, NASA; 2008a. Available from http://ston.jsc.nasa.gov/collections/trs/_techrep/TM-2008-214775Vol2.pdf

Bessone L, Coffey E, Inoue N, Gittens M, Mukai C, O'Connor S, Tomi L, Ren V, Schmidt L, Sipes W, Vander Ark S, Vassin A. International space station human behavior & performance competency model, volume I. Mission Operations Directorate (MOD) International Training Control Board (ITCB) HBP Training Working Group, NASA; 2008b. Available from http://ston.jsc.nasa.gov/collections/trs/_techrep/TM-2008-214775Vol1.pdf

Bishop SL. From earth analogs to space: learning how to boldly go. In: Vakoch DA, editor. On orbit and beyond: psychological perspectives on human spaceflight. New York: Springer; 2013. p. 25–50.

Blass T. The man who shocked the world: the life and legacy of Stanley Milgram. New York: Basic Books; 2004.

Blum B. The life span of a lie. Medium; Jun 7, 2018. Available from https://medium.com/s/trustissues/the-lifespan-of-a-lie-d869212b1f62

Bois JM, Legrand G, Matsakis Y, Venet M, McIntyre J, Shulenin A. Technical evolutions of the French multipurpose instruments for cognitive neuro-sciences. Acta Astronaut. 1998;42:89–98.

Bonting S, editor. Advances in space biology and medicine, vol. 3: European isolation and confinement study. London: JAI; 1993.

Bower JL, Laughlin MS, Connaboy C, Simpson RJ, Alfano CA. Factor structure and validation of the mental health checklist (MHCL) for use in isolated, confined, and extreme environments. Acta Astronaut. 2019;161:405–14.

Brewer MB, Crano WD. Research design and issues of validity. In: Reis HT, Judd CM, editors. Handbook of research methods in social and personality psychology. 2nd ed. New York: Cambridge University Press; 2014. p. 11–26.
Buckey Jr. JC. Space physiology. Oxford: Oxford University Press; 2006.
Buckey Jr. JC, Homick JL, editors. The Neurolab Spacelab Mission: neuroscience research in space: results from the sts-90. Neurolab Spacelab Mission; 2003. Available from https://lsda.jsc.nasa.gov/refs/Neurolab/SP-2003-535.pdf
Burger JM. Replicating milgram: would people still obey today? Am Psychol. 2009;64(1):1–11.
Carnahan T, McFarland S. Revisiting the Stanford prison experiment: could participant self-selection have led to the cruelty? Pers Soc Psychol Bull. 2007;33(5):603–14.
Citi L, Tonet O, Marinelli M. Matching brain-machine interface performance to space applications. Int Rev Neurobiol. 2009;86:199–212.
Clark JB, Bacal K. Neurologic concerns. In: Barratt MR, Pool SL, editors. Principles of clinical medicine for space flight. New York: Springer; 2008. p. 361–80.
Clément G. Fundamentals of space medicine. 2nd ed. New York: Springer; 2011.
Clément G, Reschke MF. Neuroscience in space. New York: Springer; 2008.
Coffey EBJ, Brouwer AM, Wilschut ES, van Erp JBF. Brain–machine interfaces in space: using spontaneous rather than intentionally generated brain signals. Acta Astronaut. 2010;67:1–11.
De la Torre GG. Cognitive neuroscience in space. Life. 2014;4:281–94.
Delp MD, Charvat JM, Limoli CL, Globus RK, Ghosh P. Apollo lunar astronauts show higher cardiovascular disease mortality: possible deep space radiation effects on the vascular endothelium. Sci Rep. 2016;6:29901.
Dille JR, Mohler SR. The beginnings: past and present. In: Davis JR, Johnson R, Stepanek J, Fogarty JA, editors. Fundamentals of space medicine. 4th ed. Philadelphia: Lippincott Williams and Wilkins; 2008. p. 1–19.
Doris JM. Lack of character: personality and moral behavior. New York: Cambridge University Press; 2002.
Doris JM, The Moral Psychology Research Group. The moral psychology handbook. Oxford: Oxford University Press; 2010.
Flynn CF. Behavioral health and performance support. In: Barratt MR, Pool SL, editors. Principles of clinical medicine for space flight. New York: Springer; 2008. p. 391–412.
Franzen MD, Wilhelm KL. Conceptual foundations of ecological validity in neuropsychological assessment. In: Sbordone RJ, Long CJ, editors. Ecological validity of neuropsychological testing. Delray Beach: GR/St. Lucie; 1996. p. 91–112.
Furukawa S. How to train your mind: lessons from astronauts (In Japanese). Tokyo: MyNavi Books; 2013.
Gray P. Why Zimbardo's prison experiment isn't in my textbook. Psychol Today; Oct 19, 2013. https://www.psychologytoday.com/us/blog/freedom-learn/201310/why-zimbardo-s-prison-experiment-isn-t-in-my-textbook
Hadfield C. An astronaut's guide to life on earth: what going to space taught me about ingenuity, determination, and being prepared for anything. New York: Little, Brown; 2013.
Haney G, Banks C, Zimbardo P. Interpersonal dynamics in a simulated prison. Int J Criminol Penol. 1973;1:69–97.
Hariri AR, Holmes A. Finding translation in stress research. Nat Neurosci. 2015;18(10):1347–52.
Harrison AA. Behavioral health: integrating research and application in support of exploration missions. Aviat Space Environ Med. 2005;76(6 Suppl):B3–B12.
Harrison AA, Fiedler ER. Introduction: psychology and space exploration. In: Vakoch DA, editor. On orbit and beyond: psychological perspectives on human space-flight. New York: Springer; 2013a. p. xxvii–xl.
Harrison AA, Fiedler ER. Behavioral health. In: Vakoch DA, editor. On orbit and beyond: psychological perspectives on human spaceflight. New York: Springer; 2013b. p. 3–24.
Harrison AA, Clearwater YA, McKay CP. From Antarctica to outer space: life in isolation and confinement. New York: Springer; 1991.

Hartman E. The virtue approach to business ethics. In: Russell DC, editor. The Cambridge companion to virtue ethics. New York: Cambridge University Press; 2013. p. 240–64.

Houts A, Gholson B. Brownian notions: one historical philosopher's resistance to psychology of science via three truism and ecological validity. Soc Epistemol. 1989;3(2):139–46.

International Space Life Science Working Group. Space life sciences flight experiments information package; 2014. Available from http://iss.jaxa.jp/kiboexp/participation/application/documents/life_med_2014/feip2014.pdf

Jacobsen A. The Pentagon's brain: an uncensored history of DARPA, America's top-secret military research agency. New York: Little, Brown; 2015.

Jeong H, Yim HW, Song YJ, Ki M, Min JA, Cho J, et al. Mental health status of people isolated due to Middle East Respiratory Syndrome. Epidemiol Health. 2016;38:e2016048.

Kanas N, Manzey D. Space psychology and psychiatry. 2nd ed. New York: Springer; 2008.

Kanas N, Sandal G, Boyd JE, Gushin VI, Manzey D, North R, Leon GR, Suedfield P, Bishop S, Fiedler ER, Inoue N, Johannes B, Kealey DJ, Kraft N, Matsuzaki I, Musson D, Palinkas LA, Salnitskiy VP, Sipes W, Stuster J, Wang J. Psychology and culture during long-duration space mission. Acta Astronaut. 2009;64:659–77.

Kovaleva AA, Skedina MA, Potapov MG. Human neurophysiological state in long-term isolation. Hum Physiol. 2017;43(7):741–7.

Kuznetsova PG, Gushchin VI, Vinokhodova AG, Chekalina AI, Shved DM. Interpersonal interaction under the conditions of high autonomy in interplanetary mission simulation (Mars-500 Experiment). Hum Physiol. 2017;43(7):751–6.

Le Texier T. Debunking the Stanford prison experiment. Am Psychol. 2019; https://doi.org/10.1037/amp0000401.

Mallis MM, DeRoshia CW. Circadian rhythms, sleep, and performance in space. Aviat Space Environ Med. 2005;76(6 Suppl):B94–107.

Maunder R, Hunter J, Vincent L, Bennett J, Peladeau N, Leszcz M, Sadavoy J, Verhaeghe LM, Steinberg R, Mazzulli T. The immediate psychological and occupational impact of the 2003 SARS outbreak in a teaching hospital. Can Med Assoc J. 2003;168(10):1245–51.

McCoy AW. A question of torture: CIA interrogation, from the cold war to the war on terror. New York: Metropolitan Books; 2006.

Mehlman MJ. Captain America and Iron Man: biological, genetic, and psychological enhancement and the warrior ethos. In: Lucas G, editor. Routledge handbook of military ethics. London; 2015. p. 406–20.

Milgram S. Behavioral study of obedience. J Abnorm Soc Psychol. 1963;67:371–8.

Milgram S. Obedience to authority: an experimental view. New York: Harper and Row; 1974.

Millán JR, Ferrez PW, Buttfield A. Non invasive brain-machine interfaces: final report. Paris: ESA; 2007. Available from https://www.esa.int/gsp/ACT/doc/ARI/ARI%20Study%20Report/ACT-RPT-BIO-ARI-056402-Non_invasive_brain-machine_interfaces_-_Pisa_E_Piaggio.pdf

Mitchell ML, Jolley JL. Research design explained. 5th ed. Berkeley, CA: Thomson/Wadsworth Learning; 2004.

National Aeronautics and Space Administration. The brain in space: a teacher's guide with activities for neuroscience. Washington, DC: NASA; 2000. Available from https://www.nasa.gov/pdf/58294main_The.Brain.in.Space.pdf

National Aeronautics and Space Administration. International space station: environmental control and life support system; 2008. Available from https://www.nasa.gov/sites/default/files/104840main_eclss.pdf

National Aeronautics and Space Administration. NASA's response to mine disaster remembered; 2011. Available from http://www.nasa.gov/news/chile_assistance.html

National Aeronautics and Space Administration. Environmental control and life support system (ECLSS); 2017. Available from https://www.nasa.gov/sites/default/files/atoms/files/g-281237_eclss_0.pdf

Naval Aerospace Medical Institute. U.S. naval: flight surgeon's manual. 3rd ed. Washington, DC: U.S. Navy; 1991.

Nicogossian AE, Williams RS, Huntoon CL, Doarn CR, Polk JD, Schneider VS. Space physiology and medicine: from evidence to practice. 4th ed. New York: Springer; 2016.

Open Science Collaboration. Estimating the reproducibility of psychological science. Science. 2015;349:943.

Parihar VK, Allen B, Tran KK, Macaraeg TG, Chu EM, Kwok SF, Chmielewski NN, Craver BM, Baulch JE, Acharya MM, Cucinotta FA, Limoli CL. What happens to your brain on the way to Mars. Sci Adv. 2015;1:e1400256.

Resnick B. The Stanford prison experiment was massively influential: we just learned it was a fraud. Vox; Jun 13, 2018. https://www.vox.com/2018/6/13/17449118/stanford-prison-experiment-fraud-psychology-replication

Roberts DR, Albrecht MH, Collins HR, Asemani D, Chatterjee AR, Spampinato MV, Zhu X, Chimowitz MI, Antonucci MU. Effects of spaceflight on astronaut brain structure as indicated on MRI. N Engl J Med. 2017;377(18):1746–53.

Salamon N, Grimm JM, Horack JM, Newton EK. Application of virtual reality for crew mental health in extended-duration space missions. Acta Astronaut. 2018;146:117–22.

Salvador R, Folger R. Business ethics and the brain. Bus Ethics Q. 2009;19(1):1–30.

Sbordone RJ. Ecological validity: some critical issues for the neuropsychologist. In: Sbordone RJ, Long CJ, editors. Ecological validity of neuropsychological testing. Delray Beach, FL: St. Lucie Press/GR Press; 1996. p. 15–41.

Schneider S, Brümmer V, Carnahan H, Dubrowski A, Askew CD, Strüder HK. What happens to the brain in weightlessness? A first approach by EEG tomography. Neuroimage. 2008;42:1316–23.

Schneider S, Brümmer V, Carnahan H, Kleinert J, Piacentini MF, Meeusen R, Strüder HK. Exercise as a countermeasure to psycho-physiological deconditioning during long-term confinement. Behav Brain Res. 2010;211:208–14.

Searfoss R. Liftoff: an astronaut commander's countdown for purpose powered leadership. New York: Morgan James; 2016.

Seidler RD, Reuter-Lorenz P, Mulavara AP, Bloomberg JJ, Wood SJ. Spaceflight effects on neurocognitive performance: extent, longevity, and neural bases; 2018. Available from https://www.nasa.gov/mission_pages/station/research/experiments/explorer/Investigation.html?#id=979

Shadish WR, Cook TD, Campbell DT. Experimental and quasi-experimental designs for generalized causal inference. Boston: Houghton Mifflin; 2002.

Shved DM, Gushchin VI, Vinokhodova AG, Nichiporuk IA, Vasilieva GY. Psychophysiological adaptation and communication behavior of a human operator during 105-day isolation. Hum Physiol. 2014;40(7):797–803.

Simpson JA, Rholes WS, Nelligan JS. Support seeking and support giving within couples in an anxiety-provoking situation: the role of attachment styles. J Pers Soc Psychol. 1992;62 (3):434–46.

Slack KJ, Williams TJ, Schneiderman JS, Whitmire AM, Picano JJ. Evidence report: risk of adverse cognitive or behavioral conditions and psychiatric disorders. Houston: NASA; 2016. Available from https://humanresearchroadmap.nasa.gov/evidence/reports/bmed.pdf

Sreenivasan G. The situationist critique of virtue ethics. In: Russell DC, editor. The Cambridge companion to virtue ethics. New York: Cambridge University Press; 2013. p. 290–314.

Stuster J. Behavioral issues associated with long duration space expeditions: review and analysis of astronaut journals experiment 01-E104 (journals), phase 2 final report. Houston: NASA; 2016. Available from https://ston.jsc.nasa.gov/collections/TRS/_techrep/TM-2016-218603.pdf

Sulzman FM, Wolfe JW. Neurosciences research in space: future directions. Acta Astronaut. 1991;23:289–93.

Tachibana K. Moral neuroscience and moral philosophy: interactions for ecological validity. Kagaku Tetsugaku (J Philos Sci Soc Jpn). 2009;42(2):41–58.

Tachibana K. Space neuroethics. Am J Bioethics: Neurosci. 2017;8(1):W6–W7.

Tachibana S, Sano S, Inoue N. Stress factors among flight passengers who have been isolated and confined in quarantine due to the containment operations for Influenza (H1N1) 2009:

comparison with similar isolation and confinement environments (In Japanese). J Natl Defense Medical College. 2012;37(1):27–38.

Tachibana K, Tachibana S, Inoue N. From outer space to Earth—the social significance of isolated and confined environment research in human space exploration. Acta Astronaut. 2017;140:273–83.

Tafforin C. Time effects, cultural influences, and individual differences in crew behavior during the Mars-500 experiment. Aviat Space Environ Med. 2013;84(10):1082–6.

Tafforin C, Abati FG. Cultural ethology as a new approach of interplanetary crew's behavior. Acta Astronaut. 2017;139:102–10.

Ursin H, Comet B, Soulez-Larivière C. An attempt to determine the ideal psychological profiles for crews of long term space missions. Adv Space Res. 1991;12(1):301–14.

Vinokhodova AG, Gushchin VI, Eskov KN, Khananashvili MM. Psychological selection and optimization of interpersonal relationships in an experiment with 105-days isolation. Hum Physiol. 2012;38(7):677–82.

Wakata K. Instant judgments: seven skills that I improved as an astronaut (In Japanese). Tokyo Nippon Jitsugyo Publishing; 2016.

Wolfe T. The right stuff. New York: Farrar, Straus and Giroux; 1979.

Wotring EW. Space pharmacology. New York: Springer; 2012.

Yamaguchi T. Adoption criteria of astronauts: how to measure leadership (In Japanese). Tokyo: Kadokawa Shoten; 2014.

Young LR. Vestibular reactions to spaceflight: human factors issues. Aviat Space Environ Med. 2000;71(9 Suppl):A100–4.

Zange J, Müller K, Schuber M, Wackerhage H, Hoffmann U, Günther RW, Adam G, Neuerburg JM, Sinitsyn VE, Bacharev AO, Belichenko OI. Changes in calf muscle performance, energy metabolism, and muscle volume caused by long term stay on space station MIR. Int J Sports Med. 1997;18(4):S308–S9.

Zimbardo P. The Lucifer effect: understanding how good people turn evil. New York: Random House; 2007.

Conclusion

The world of organizational science has evolved tremendously in the last decades with increasingly specific methodological and theoretical developments with respect to all aspects of individual behavior in organizations as well as broader group-based theories and methodologies. Like for other areas of scholarship in the humanities and the social sciences, the possibility that neuroscience provides additional insights into human behavior and increasingly powerful means of interventions therein has seized the attention of academics and those in the field. Organizational neuroscience crystallizes these hopes and aspirations. At this time, the true scientific and practical contribution of organizational neuroscience are debated. Nonetheless, it is important to start reflecting on how the knowledge generated by organizational neuroscience intersects with situations where human values are at stake (e.g., the privacy of employees, respect for the agency and autonomy of employees, corporate responsibility). It is possible that insights into human behavior in organizational contexts yield knowledge conducive of human betterment and that, accordingly, what is learnt will be used to improve life conditions and wellness. But it is also true that sharper manipulation of human behaviors, attitudes, and emotions is entirely possible. Nothing is predetermined in this respect, and the ethical use of organizational neuroscience is not intrinsic to its development. This observation served as an impetus for this book and hopefully for further reflection on ethical aspects of organizational neuroscience.

We hope readers have appreciated this first volume dedicated to the topic of organizational neuroethics. We hope that the contributions in this book spark conversations on the developments, promises, ethical issues, and limitations associated with organizational neuroscience. This area of research and its derived applications are developing rapidly. Without undermining needlessly this enthusiasm, there is still a need to adopt a precautious and interrogative stance. We encourage readers to question assumptions and applications of organizational neuroscience and to consider them through the prism of human values.

J. T. Martineau, E. Racine (eds.), *Organizational Neuroethics*, Advances in Neuroethics, https://doi.org/10.1007/978-3-030-27177-0

MIX
Papier aus verantwortungsvollen Quellen
Paper from responsible sources
FSC® C105338

If you have any concerns about our products,
you can contact us on
ProductSafety@springernature.com

In case Publisher is established outside the EU,
the EU authorized representative is:
Springer Nature Customer Service Center GmbH
Europaplatz 3, 69115 Heidelberg, Germany

Printed by Libri Plureos GmbH
in Hamburg, Germany